Thermodynamic & Transport Properties of Fluids

by D. James Benton

Foreword

Accurate, consistent, and continuous thermodynamic and transport properties are essential to the analysis and design of energy devices of all sorts, from power generation to product manufacturing. Articles and papers abound covering various aspects of this important field. Often these are esoteric and omit details on how the process is accomplished. The end result of property research may be inaccessible to practitioners, who would use the information to create and manage the machines of industry. This text is a step-by-step manual on why and how to develop and implement functions for thermodynamic and transport properties from raw data to Excel® Add-Ins.

All of the examples contained in this book,
(as well as a lot of free programs) are available at...

https://www.dudleybenton.altervista.org/software/index.html

In this text we will cover all of the significant equations of state, their strengths and weaknesses. We will also cover several transport property formulations and why these have been used. This text includes a unique graph never before published, which is perhaps the most revealing presentation of an equation of state, including the elusive meta-stable states under the vapor dome.

Programming

All of the examples in this book are implemented in the C programming language, including the Excel® Add-Ins. Several spreadsheets are also provided illustrating the use and form of the various properties. A suite of related programs (with source code) as well as useful fluid property data are freely available at the location above.

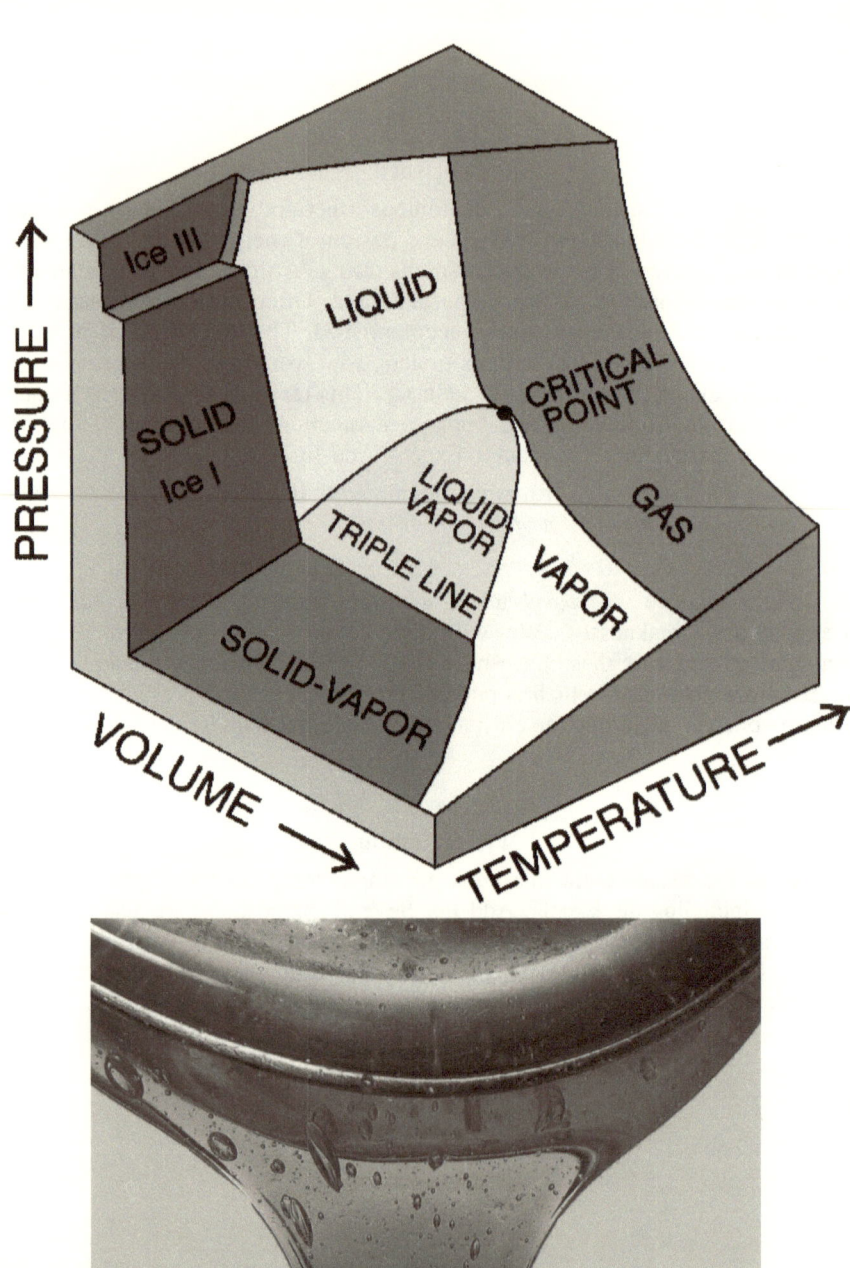

ii

Table of Contents

Chapter 1. Simple Equations of State

Pressure, (specific) Volume, Temperature (PVT) data is the most basic information we must have for any fluid. Density (ρ) is mass per unit volume or the reciprocal of specific volume. This trio is sometimes written in terms of density, PρT. Transport properties, such as viscosity and thermal conductivity, are either derived from or described in terms of these more basic quantities. We will only consider fluids, that is, gases and liquids. We will not limit our consideration to strictly Newtonian fluids (i.e., ones exhibiting a linear relationship between stress and strain), but we will not consider substances that can sustain a finite shear without moving (e.g., mayonnaise, ketchup, peanut butter, tar, glue, etc.), as most of these fail the following criterion anyway.

All such fluids that are chemically stable (i.e., their molecular structure does not change over a sufficient range of temperature and pressure), exhibit saturation behavior and a critical point. In particular, these form a vapor dome, the peak of which is the critical point. This means that the liquid and vapor can coexist in equilibrium over a range of temperatures and pressures. The highest temperature and pressure at which this behavior occurs is the critical point, at which the liquid and vapor are indistinguishable (i.e., same density and specific energy). Many substances break down chemically or change structurally (e.g., differing inter-atomic bonds) so that they never reach a critical point. We will not consider these. We will also not consider solids or the triple point (where solid, liquid, and vapor coexist in equilibrium), other than as a lower limit.

Temperature and pressure are independent, that is, these two describe the state of the fluid—except at saturation, when the liquid and vapor coexist in equilibrium. In these conditions, the temperature and pressure are related and no longer independent. Temperature and density (or specific volume) or pressure and density (or specific volume) are always independent. We will formulate all thermodynamic properties in terms of temperature and density. While we could formulate properties in terms of pressure and density, this is most inefficient and rather ineffective. We will not formulate properties in terms of pressure and temperature, as such would be quite foolish.[1]

van der Waals EOS

An equation of state (EOS) is an expression that relates pressure, density, and temperature for a fluid. The first EOS to exhibit the behavior described

---

[1] One might argue that there was adequate motivation for such foolishness in 1967, before the invention of microcomputers, but there was certainly no valid excuse for it in 1997, on the doorstep of the twenty-first century. In 2019 this would be equivalent to providing a stick and carrot with every new car so that the driver could coax it down the road like a medieval donkey cart.

1

above was introduced by van der Waals in 1873.[2] The most common form is Equation 1.1:

$$P = \frac{RT}{V-b} - \frac{a}{V^2} \tag{1.1}$$

Here P is pressure, R is the ideal gas constant, T is absolute temperature, V is the specific volume ($1/\rho$), a and b are constants specific to a particular fluid. This is a modification of the ideal gas law (Equation 1.2), which was first introduced by Émile Clapeyron in 1834.[3]

$$Z = \frac{PV}{RT} = 1 \tag{1.2}$$

R is the same in both equations. Z is the compressibility, which is unity for an ideal gas. There are two common ways of determining the constants a and b in Equation 1.1: first, partial derivatives at the critical point and 2) best fit of actual data. There is no point arguing for either one, as the equation does not accurately describe any known fluid. It does, however, have the correct general shape, which is why it is a valuable consideration. We will first plot the equation, then see how the constants might be defined.

Before plotting Equation 1.1, we will introduce some relationships that generalize while non-dimensionalizing. These are the reduced pressure, specific volume, and temperature. The subscript R indicates *reduced* and the subscript C indicates *critical*.

$$P_R = \frac{P}{P_C}$$
$$V_R = \frac{V}{V_C} \tag{1.3}$$
$$T_R = \frac{T}{T_C}$$

Introducing these non-dimensional quantities into Equation 1.1 yields:

$$Z_C P_R = \frac{T_R}{V_R - B} - \frac{A}{V_R^2} \tag{1.4}$$

---

[2] Johannes Diderik van der Waals (1837–1923) Dutch theoretical physicist and thermodynamicist.
[3] Benoît Paul Émile Clapeyron (1799–1864) French engineer and physicist.

In Equation 1.4, Z_C is the critical compressibility (i.e., Equation 1.2 with the critical values). Modified parameters $B=b/V_C$ and $A=a/V_C/R/T_C$ are non-dimensional.

The following figure is the most common representation of the van der Waals EOS. The critical point is shown by a red asterisk. The red curve to the left of this point is the locus of all saturated liquids and the red curve to the right of this point is the locus of all saturated vapors. The red curves form what is called the *vapor dome*, with the critical point at its apex.

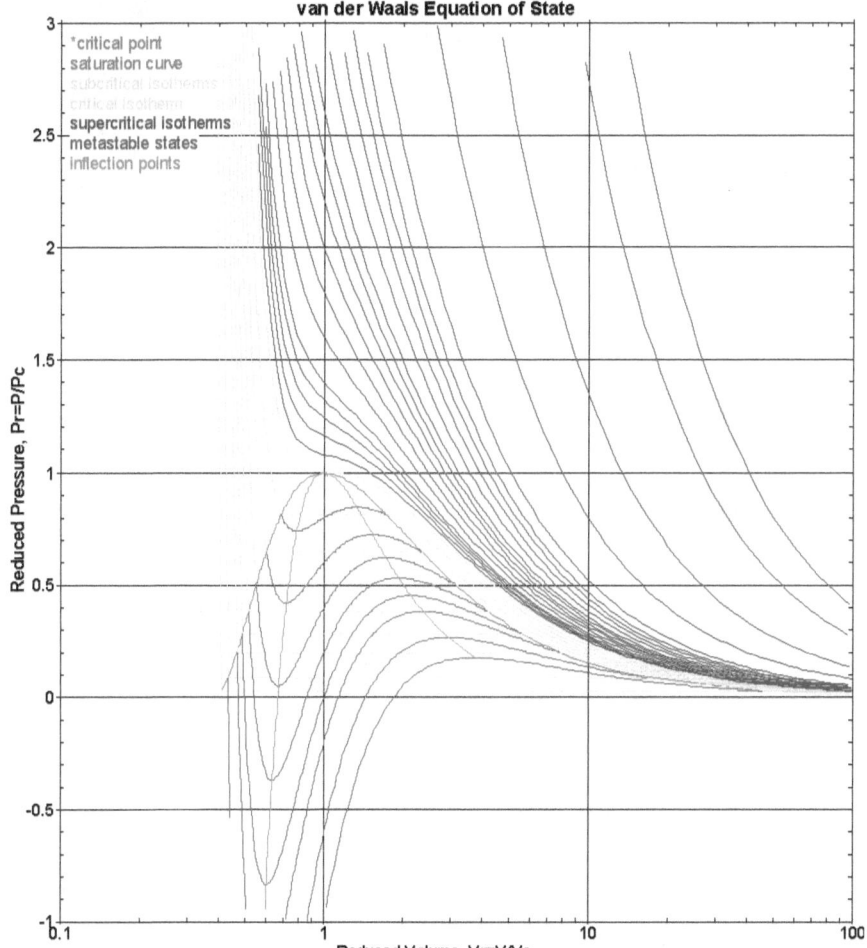

The green and blue curves are isotherms (i.e., lines of constant temperature). The green ones are for temperatures below the critical point and the blue ones are for temperatures above the critical point. The cyan curve is the

3

critical isotherm. Notice that none of the green, blue, or cyan curves ever cross. This will be an important qualification of any EOS we consider.

The horizontal green segments crossing from the red curve on the left to the red curve on the right show the range of equilibrium specific volumes (or densities) from the saturated liquid state to the saturated vapor state. The brown curves are meta-stable states, that is, they are not in equilibrium and cannot be maintained. At some time, you may have heated a liquid (perhaps coffee) on the stove or in the microwave oven to a temperature above the boiling point, yet it didn't boil until you stirred it or sprinkled something into it like sweetener or creamer. Then it suddenly boiled up, even frothed over the top. The liquid was in a meta-stable state. When you disturbed it, the liquid rapidly shifted to the equilibrium state, which meant some of liquid became vapor. It jumped from a brown curve to a green line.

You may also have touched the calm surface of very cold water, only to have it quickly freeze, sending ice crystals out from the point of contact. This is the same phenomenon—rapid shifting from a meta-stable state to equilibrium. In the first case, the liquid was at first slightly hotter than the boiling point (i.e., saturation temperature). In the second case, the liquid was at first slightly colder than the freezing point (also saturation temperature). The green-to-brown-to-green curves are continuous (i.e., unbroken slope); whereas, the green horizontal lines between the two red saturation curves are discontinuous in slope with respect to the green curves. The discontinuity is caused by a change of phase (liquid to vapor or vice versa).

The hottest you can possibly warm a liquid from the left is where the brown curve bottoms out. The coldest you can possibly cool a vapor from the right is where the brown curve tops out. The magenta curve goes through these inflections, which are called *spinodal* points. These represent the theoretical maximum departure from equilibrium. As mentioned before, the van der Waals equation is not very accurate, which is why it's even more surprising that it exhibits all these important features of real fluids.

Consider the following expanded view near the critical point:

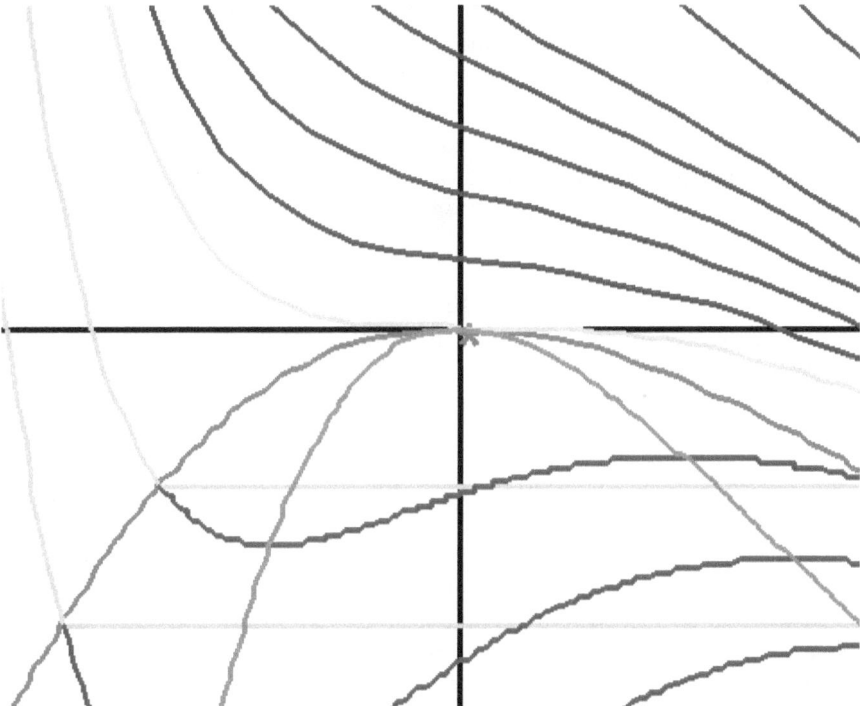

The critical isotherm (cyan curve) is flat where it crosses $P_R=T_R=1$ (the red asterisk at the critical point). In order to be flat, both the first and second partial derivative with respect to reduced specific volume (or density) must be zero. This can be stated:

$$\frac{\partial P}{\partial V} = \frac{\partial^2 P}{\partial V^2} = 0 \qquad (1.5)$$

Combining Equation 1.4 and 1.5 yields the following result:

$$A = \frac{9}{8}$$
$$B = \frac{1}{3} \qquad (1.6)$$

Plugging these two values back into Equation 1.4 yields:

$$Z_C = \frac{3}{8} \qquad (1.7)$$

Water has one of the lowest critical compressibility factors (0.2333) and Helium has one of the highest (0.3297). No known substance has a value near 0.375. This is one shortcoming of the van der Waals EOS. The other is that the sub-critical isotherms (green curves on the left side) rise much too quickly.

Motivation behind the van der Waals EOS arises from two observations: the repulsive force at high densities and an attractive (cohesive) force at low densities. The repulsive force is dominated by P proportional to $1/(V-b)$ is that of hard spheres. When $V=b$, $P=\infty$. The cohesive effect (a/V^2) is called the van der Waals force.

Dieterici EOS

Insufficient accuracy of the van der Waals EOS led to the proposal of several other simple equations of state in the decades that followed its introduction. One such equation was proposed by Dieterici,[4] which may be written as Equation 1.8:

$$Z_C P_R = \frac{T_R}{\left(V_R - \dfrac{1}{2}\right) e^{\frac{2}{T_R V_R}}} \tag{1.8}$$

The Dieterici EOS is less accurate than van der Waals. The vapor dome is much too narrow, the sub-critical isobars rise too quickly, the liquid spinodal (left magenta curve) is too shallow and so is the vapor (right magenta curve).

---

[4] C. Dieterici, (German) *Annals of Physical Chemistry* (Wiedemann's Annalen der Physik und Chemie), Vol. 69, p. 685, 1899.

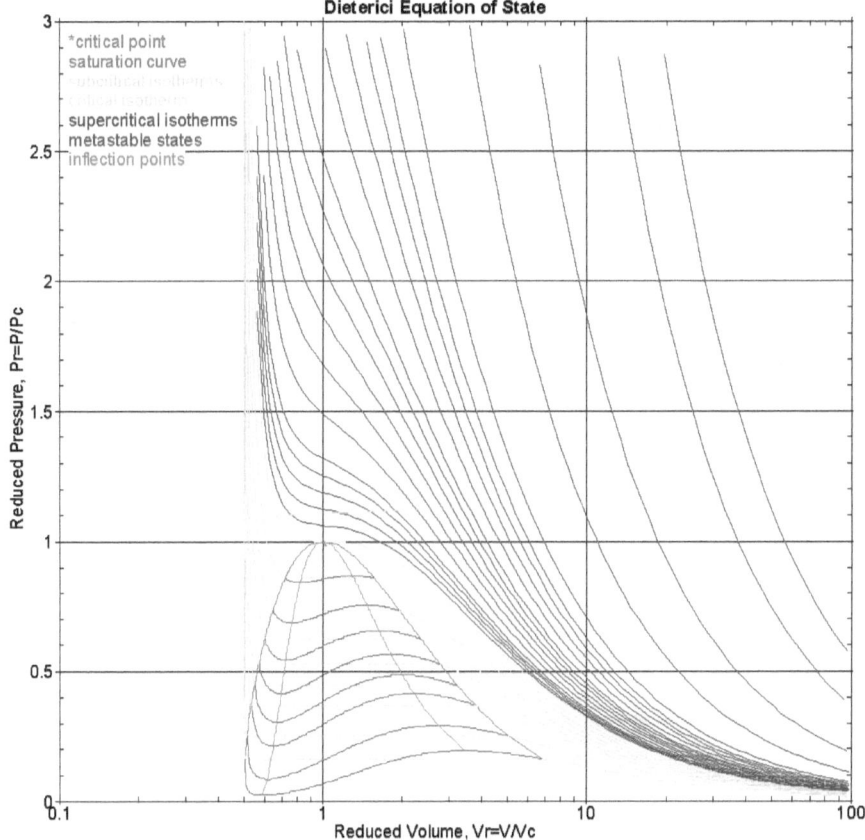

Dieterici Equation of State

- *critical point
- saturation curve
- unmodified isotherms
- critical isotherm
- supercritical isotherms
- metastable states
- inflection points

Reduced Pressure, Pr=P/Pc

Reduced Volume, Vr=V/Vc

Berthelot EOS

Berthelot[5] introduced an equation of state in 1902, which is identical to van der Waals, except for A varying with temperature.[6]

$$Z_C P_R = \frac{T_R}{(V_R - B)} - \frac{A}{T_R V_R^2} \qquad (1.9)$$

This simple modification is actually quite an improvement over the van der Waals EOS, although it isn't quite as obvious from the *PV* graph. It did, however, inspire other researchers to investigate temperature variation for both parameters *A* and *B*.

---

[5] Daniel Berthelot (1865-1927) French biologist and physicist, professor and researcher at the Academy of Sciences and the Academy of Medicine.

[6] D. Berthelot, "Travaux et Mémoires du Bureau International des Poids et Mesures," Vol. 13, 1907.

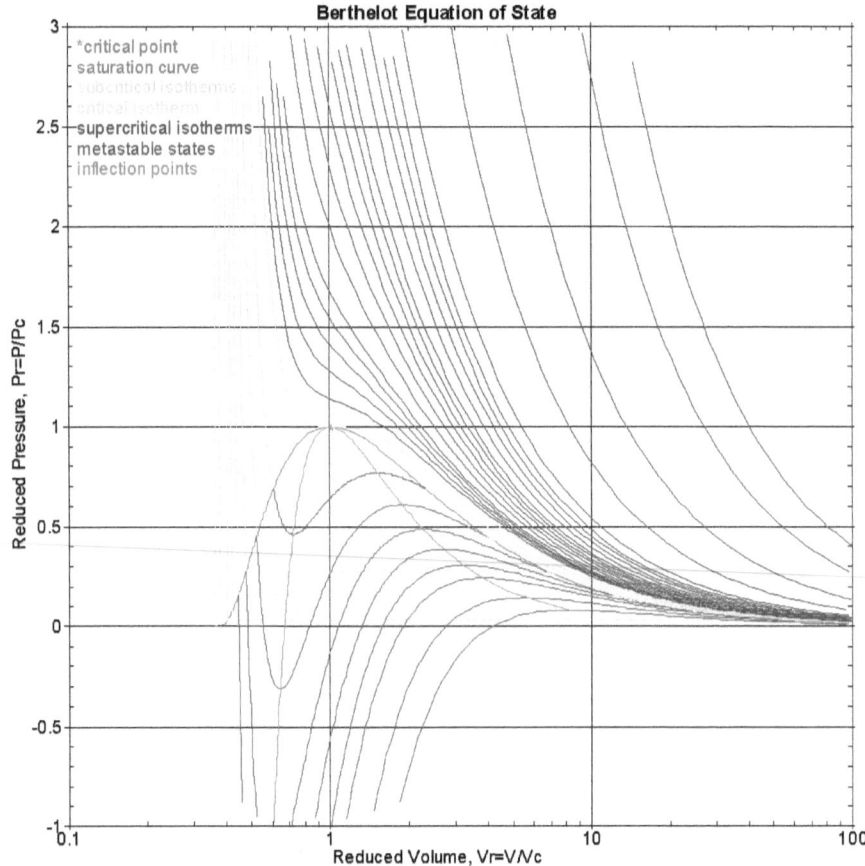

Berthelot Equation of State

Boltzmann EOS

Boltzmann[7] also introduced an equation of state, which can be expressed by Equation 1.10:

$$Z_C P_R = \frac{T_R}{V_R}\left(1 + \frac{B}{V_R} + \frac{5B^2}{V_R^2}\right) - \frac{A}{V_R^2} \qquad (1.10)$$

In this equation, $A=1+B$ and $B=1/\sqrt{15}$, resulting in $Z_C=1/3$, which is too large for most substances. This EOS is significantly less accurate than any of the three already mentioned. The vapor dome is much too large and the sub-critical isotherms do not rise nearly fast enough to approximate liquid behavior. It is

---

[7] Ludwig Eduard Boltzmann (1844–1906) Austrian physicist and philosopher, father of statistical mechanics.

only of historical interest. Boltzmann is known for his many other contributions to thermodynamics.

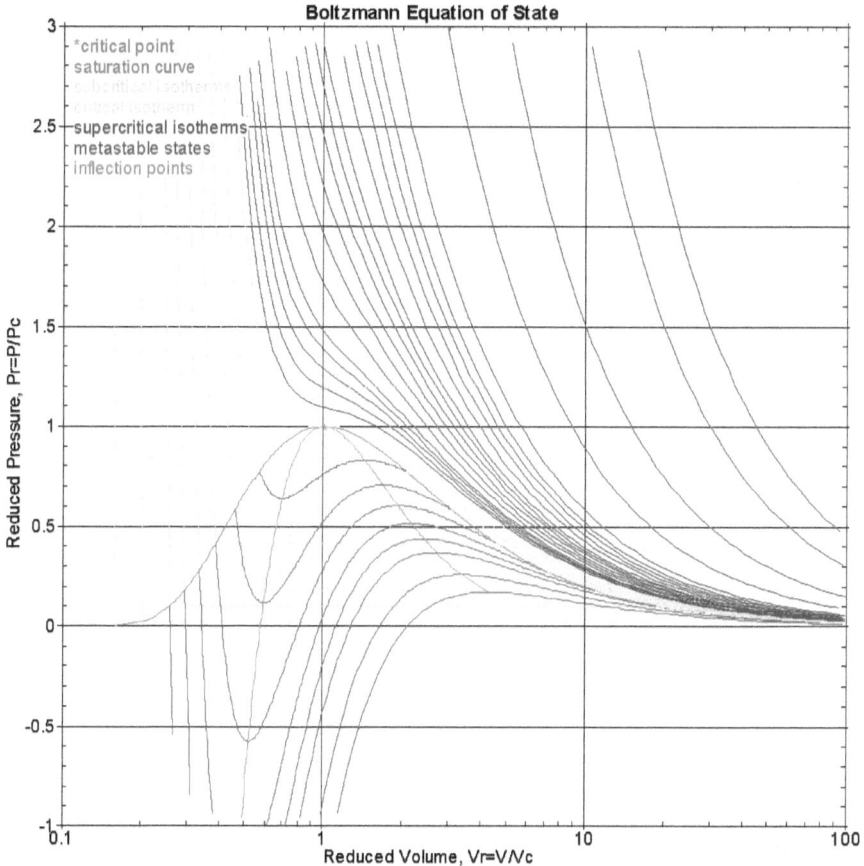

Clausius EOS

Clausius[8] introduced an equation of state that was the first improvement beyond Berthelot's, retaining the same temperature variation of parameter A and adding a term to the denominator of the van der Waals force term.

---

[8] Rudolf Julius Emanuel Clausius (1822–1888) German physicist and mathematician.

$$Z_C P_R = \frac{T_R}{(V_R - B)} - \frac{A}{T_R (V_R + C)^2}$$

$$A = \frac{(1 + C)^3}{(1 - B)^2}$$

$$C = \frac{3(1 - B)}{2} - 1 \qquad (1.11)$$

$$Z_C = \frac{1}{(1 - B)} - \frac{A}{(1 + C)^2}$$

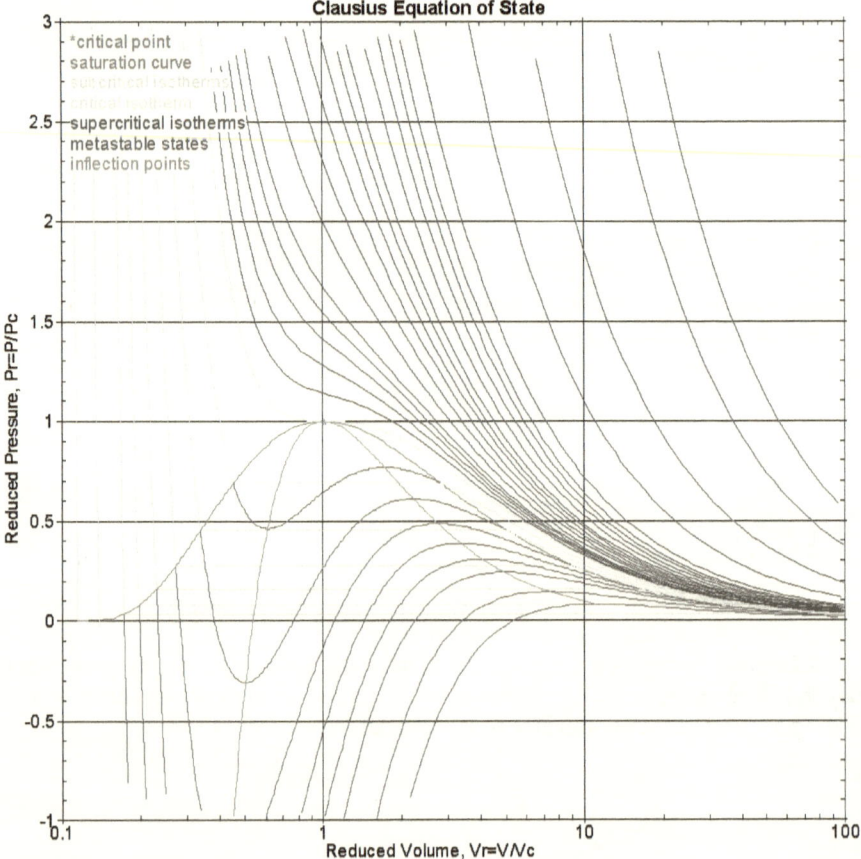

The parameter B may be adjusted for each fluid; however, Z_C must lie between 0.25 and 0.375. The vapor dome is too wide and the sub-critical isotherms are much too far apart. Still, this is the first EOS we've discussed that has some flexibility.

Redlich-Kwong EOS

While several other equations of state were introduced between van der Waals and modern times, one of the most successful was developed by Otto Redlich and Joseph Neng Shun Kwong in 1949.[9]

$$Z_C P_R = \frac{T_R}{(V_R - B)} - \frac{A}{V_R(V_R + B)\sqrt{T_R}} \qquad (1.12)$$

The parameters are somewhat empirical, with $A=0.4728/Z_C$ and $B=0.26$, making $Z_C=1/3$. The square-root variation of A with T_R is an improvement over Clausius, as is the denominator of the van der Waals force term. The critical compressibility factor is too large.

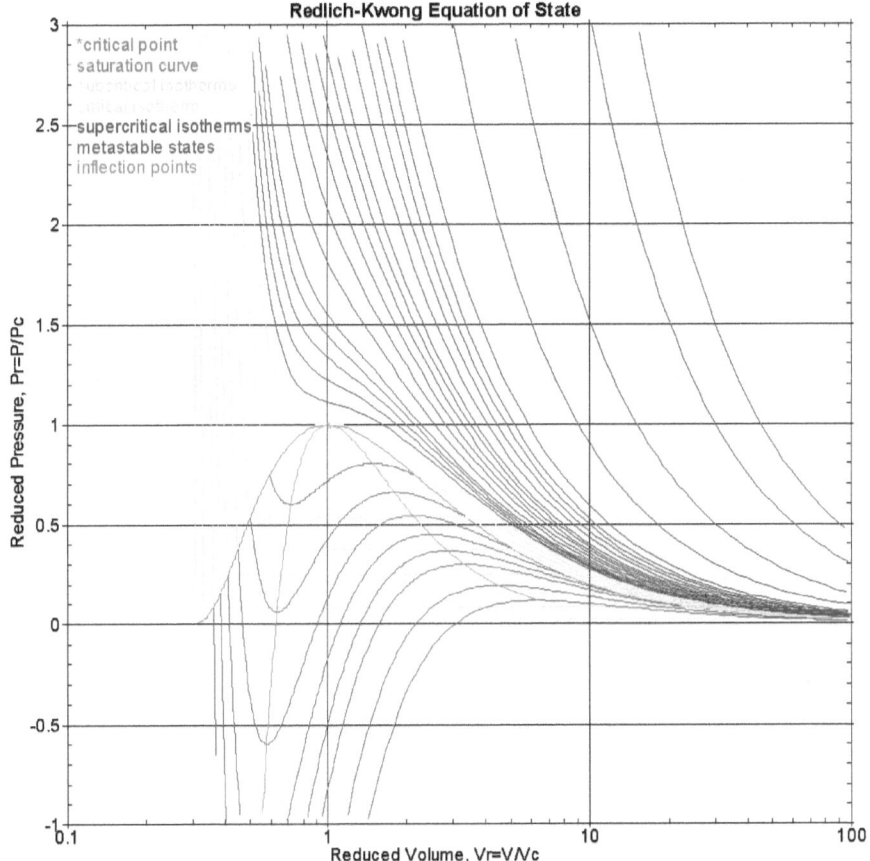

Redlich-Kwong Equation of State

[9] Redlich, O. and Kwong, J. N. S., "On The Thermodynamics of Solutions," *Chemical Review*, Vol. 44, No. 1, pp. 233–244, 1949.

Wohl EOS

The last equation of state we will consider in this section was developed by Wohl.[10]

$$Z_C P_R = \frac{T_R}{V_R - b} - \frac{a}{T_R V_R (V_R \pm b)} + \frac{c}{T_R^2 V_R^3}$$ (1.13)

The Wohl EOS is most often seen with $a/(V_R-b)$ in the second term, but with two terms having (V_R-b) in the denominator, this EOS grossly over-compensates for density. Even after changing this to (V_R+b), the discrepancies are quite large, as shown in this next figure. The Wohl EOS is of no practical use.

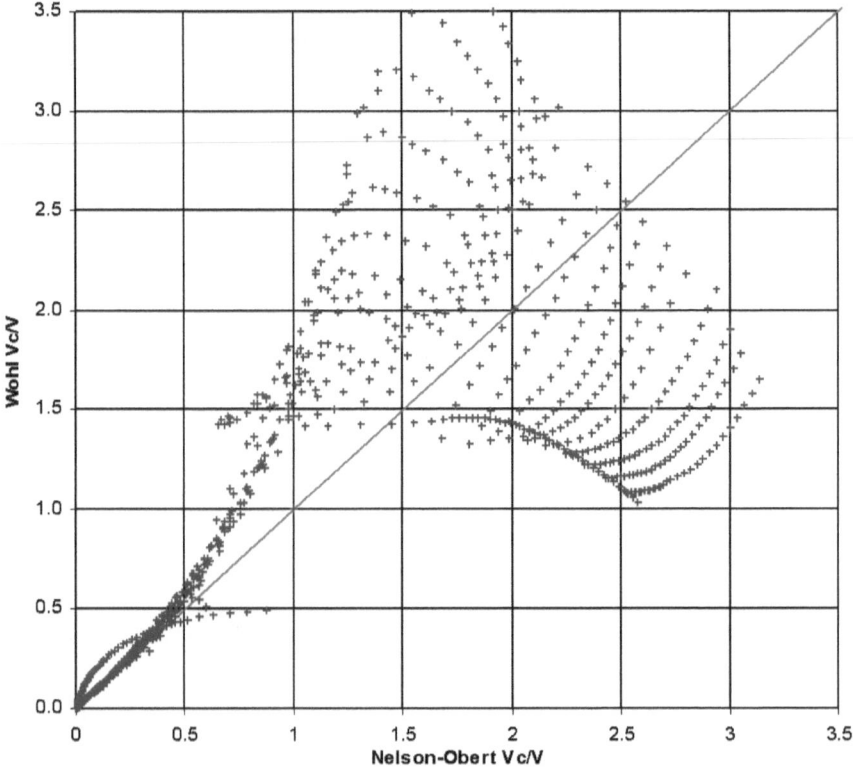

[10] Wohl, A., "Investigation of the Condition Equation", *Zeitschrift für Physikalische Chemie* (Leipzig), Vol. 87, pp. 1-39, 1914.

Cubic Equations of State

Before discussing any further improvements, we note that all of these (except for Dieterici's) are cubic, either in specific volume, density, or compressibility. For instance, the van der Waals EOS can be rearranged thus:

$$V_R^3 - \left(\frac{1}{3} + \frac{8T_R}{3P_R}\right)V_R^2 + \frac{3V_R}{P_R} - \frac{1}{P_R} = 0 \qquad (1.13)$$

We note that for sub-critical temperatures, there are three real roots (i.e., where the green curve intersect a horizontal line of constant pressure). At the critical point there are also three real roots—all the same. At temperatures above the critical ($T_R>1$, the blue curves), there is only one real root and two complex roots. The number of roots in each region will be important later in our discussion of empirical equations of state. The van der Waals EOS can also be rearranged to form a cubic equation in density:

$$\rho_R^3 - 3\rho_R^2 + \left(\frac{P_R + 8T_R}{3}\right)\rho_R - P_R = 0 \qquad (1.14)$$

The van der Waals EOS can also be rearranged to form a cubic equation in compressibility:

$$Z^3 - \left(\frac{P_R}{8T_R} + 1\right)Z^2 + \left(\frac{27P_R}{64T_R^2}\right)Z - \frac{27P_R^2}{512T_R^3} = 0 \qquad (1.15)$$

The same algebraic process can be used to transform the Berthelot, Boltzmann, Clausius, and Redlich-Kwong equations of state into equivalent expressions.

Compressibility Chart

In order to further distinguish between these and additional equations of state, in search of the most effective ones, we will now consider other representations. The first of these is the compressibility chart. The van der Waals EOS can be rearranged to form:

$$Z = \frac{V_R}{V_R - B} - \frac{A}{T_R V_R} \qquad (1.16)$$

The colors and curves represent the same isotherms, saturation curve, and vapor dome as before, only drawn on different coordinates.

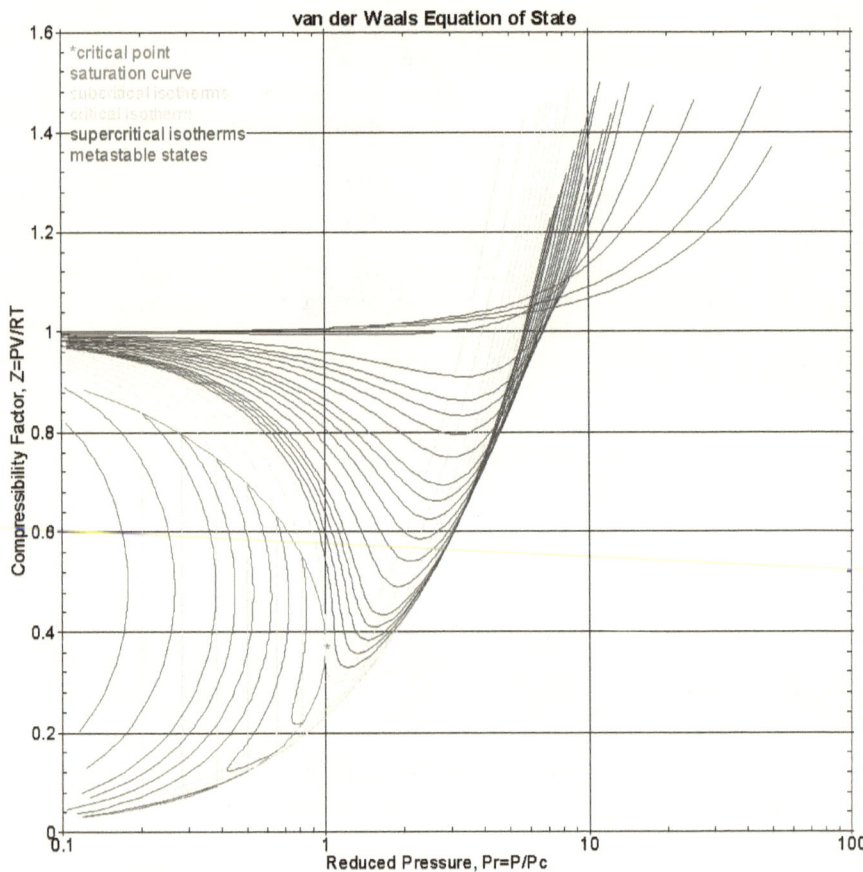

van der Waals Equation of State

Compressibility Factor, Z=PV/RT

- *critical point
- saturation curve
- supercritical isotherms
- critical isotherm
- supercritical isotherms
- metastable states

Reduced Pressure, Pr=P/Pc

One difference between the Z and PV graphs is that some of the curves cross. It is important to consider where these curves cross, if we want to accurately represent real fluid behavior.

Theorem Corresponding States

The theorem of corresponding states was first proposed by van der Waals. It is sometimes called a law or principle, but it doesn't meet the criteria for these terms.[11] In short, this theorem states that most fluids (that meet our initial requirements stated at the beginning of this chapter) will behave similarly at the same reduced pressure, temperature, and specific volume (or density). This is a formal way of saying that the PVT and Z charts should look pretty much the

---

[11] Laws must always hold, for example, conservation of energy. Principles must at least be valid most of the time, perhaps with some limited exceptions, for instance, mass is always conserved except when it's converted to energy or vice versa.

same for any such fluid if plotted on the same scales of reduced or dimensionless parameters. We've assumed that to be the case up until this point, though not stated it, in order to lay some groundwork and introduce terms.

Nelson-Obert Generalized Compressibility

Nelson and Obert gathered PVT data for dozens of common substances, omitting water, ammonia, helium, and hydrogen for reasons that we will discuss later.[12] They plotted these on graph paper the really old fashioned way and drew lines.[13] The original along with modified forms abound on the Web. These are interesting to look at, but what we really need is digitized data. You can find this in the on-line archive in the examples\Nelson-Obert folder spreadsheet Nelson-Obert.xls. There is a similar spreadsheet, van_der_Waals,xls.

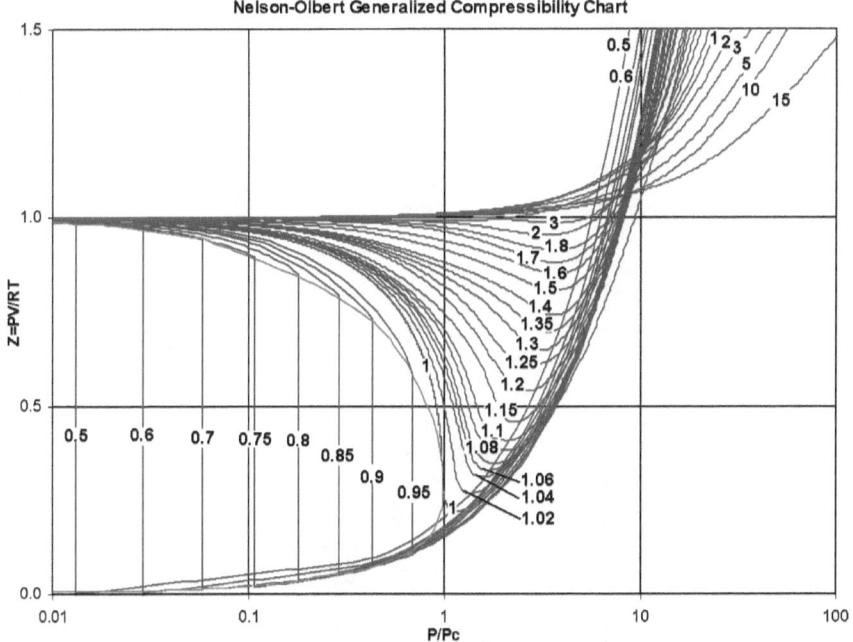

The numbers indicate reduced temperatures, T_R, so that these are isotherms. The critical point is right-most location on the red saturation curve where the $T_R=1$ isotherm just touches. The critical compressibility is set to 0.27, a

---

[12] Water and ammonia are polar compounds and exhibit very unusual properties, including very high latent and specific heats. Helium and hydrogen both have such low critical temperatures that the properties we're most interested in (e.g., the vapor dome) cover only a very small portion of the useful range.

[13] Nelson, L. C. and Obert, E. F., "Generalized PVT Properties of Gases," *Transactions of the ASME*, Vol. 76, pp. 1057–1066, 1954.

reasonable average for the substances represented. With the digitized data, we can readily calculate $V_R = ZT_R/Z_C P_R$ and produce the corresponding PVT chart as before, also included in the same spreadsheet.

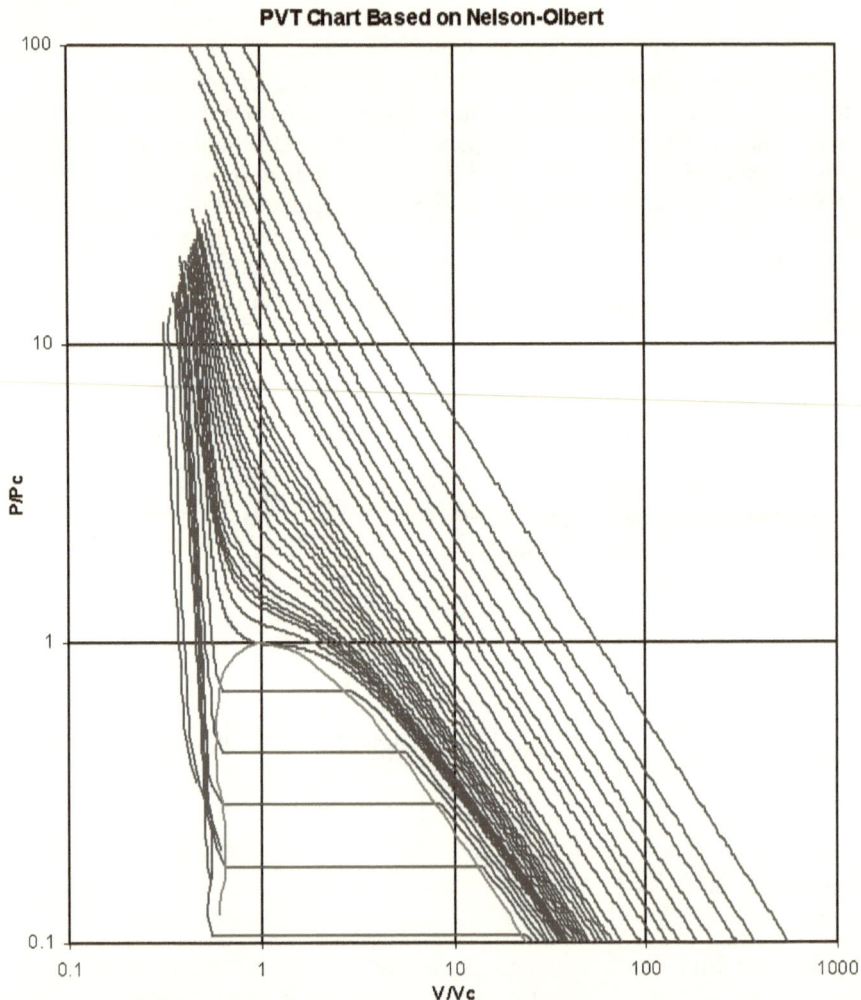

Two differences between this and the previous plot should be readily evident. First, there are no meta-stable states. Second, the sub-critical liquid region (bottom left corner) is quite "rough" and several dozen points have been eliminated. Nelson and Obert only considered equilibrium data. Meta-stable states are very difficult and problematic to measure, let alone reproduce. Also, they were not primarily interested in liquids and the compressibility chart is not

16

conducive to accurately representing liquid properties, as they all collapse to nearly the same values in this view.

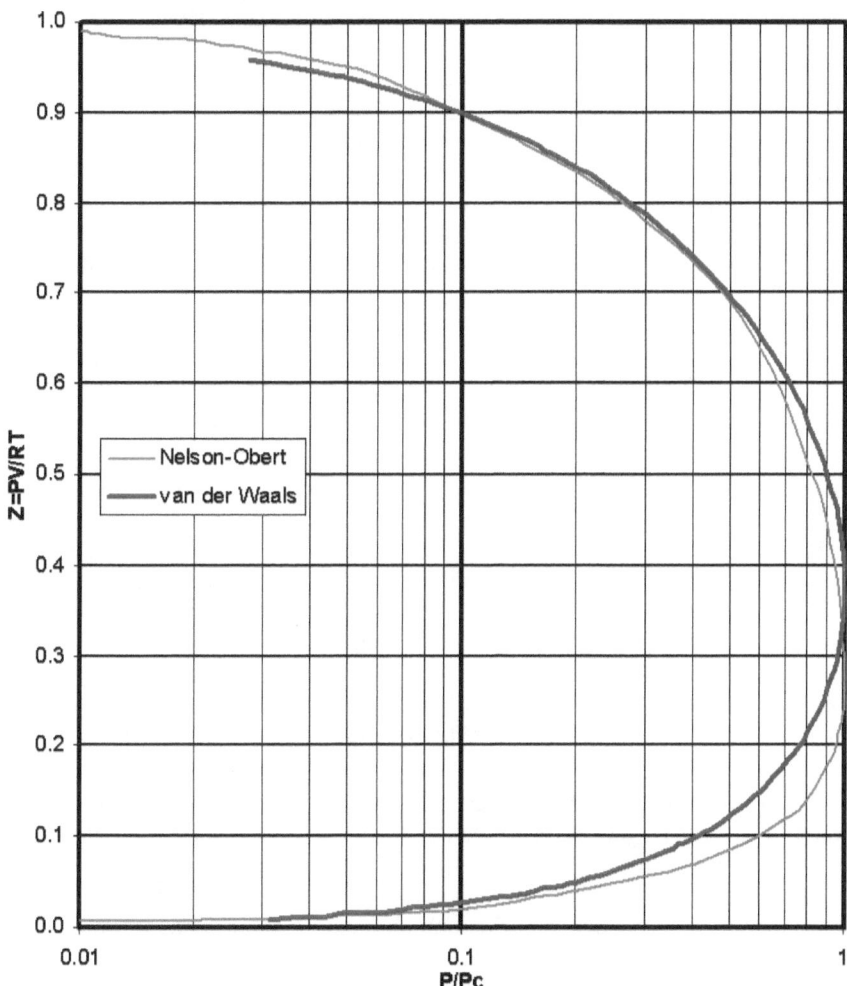

We can compare these other early and simple equations of state to the Nelson-Obert compressibility chart in order to evaluate their accuracy, as the latter is based on actual data. Perhaps the most noticeable difference between van der Waals and Nelson-Obert is the saturation curves. The (Nelson-Obert) curve is much more blunt than the theoretical (van der Waals). This means that actual liquids (the bottom part of the saturation curve) are much more dense (Z is smaller), vary less with temperature over much of the range, and rise much more rapidly to reach the critical point, which is significantly smaller. The saturated vapors (top part of the saturation curve) are fairly close, except near

17

the critical point, which is also approached more abruptly. The two saturation curves are compared in the previous figure: The shape of the saturation curve is a very important criterion for evaluating various equations of state.

Maxwell's Criterion

We haven't explained yet how the red saturation curve was obtained in the first plot of van der Waals EOS. We obtain this curve by applying Maxwell's Criterion.[14] In order to explain the origin of this very important relationship, we must first define free energy, which requires defining both entropy and enthalpy. Specific internal energy (given the symbol, u) is the energy associated with microscopic (i.e., atomic and/or molecular) motion (e.g., translation, vibration, and rotation) per unit mass, making it a macroscopic property. Specific enthalpy (given the symbol, h) is equal to the internal energy plus the product of pressure and specific volume ($h=u+pv$). Typically, internal energy is associated with closed systems (i.e., those that do not exchange mass with the surroundings) and enthalpy is associated with open systems (i.e., those that do exchange mass with the surroundings). These concepts are necessary, but we don't want to get bogged down with them at this point.

Specific entropy (given the symbol, s) was first described by Clausius (originator of the previous EOS). At this point we will only consider a very simplistic definition of entropy: an indication of the quality (more specifically, unavailability) of the total energy (either u or h, depending on the system), per unit mass, per unit change in temperature. Specific free (or available) energy is equal to the total energy less the temperature times the entropy. Two types are considered. For a closed system, the specific Helmholtz[15] free energy (abbreviated HFE and given the symbol, a) is then $a=u-Ts$. For an open system, the specific Gibbs[16] free energy (given the symbol, g) is the $g=h-Ts$.

Sorry, but this is a very long and round about way of getting to the saturation curve. In order for the liquid and vapor to be in equilibrium with each other (having the same free energy) over the entire range of conditions from saturated liquid (the left side of the saturation curve, given the subscript, f) to saturated vapor (the right side of the saturation curve, given the subscript, g), the following must hold:

$$\int_{v_f}^{v_g} pdv = p_{SAT}\left(v_g - v_f\right) \tag{1.17}$$

On a molecular level, the molecules can cross the liquid-vapor boundary in either direction with no net gain or loss of free energy. This equation also

---

[14] James Clerk Maxwell (1831–1879) Scottish scientist, physicist, and mathematician.
[15] Hermann Ludwig Ferdinand von Helmholtz (1821–1894) German physician and physicist.
[16] Josiah Willard Gibbs (1839–1903) American physicist, chemist, and mathematician.

follows from one of Maxwell's relations.[17] While this seems rather esoteric, it has a clear graphical interpretation. In the following figure, the yellow and red areas must be equal. The horizontal axis is a log scale, so they don't appear to be equal in this view, but they are. The left side of Equation 1.17 is the area under the brown curve. The right side of this equation is the area under the flat (horizontal) green line. We find each point along the saturation curve by simultaneously solving for the two roots (v_f and v_g where $p=p_{SAT}$) and the integral of pdv between these two points. This is why it's important how deep the brown curve falls on the left side and how high it rises on the right side for a particular EOS, as this leads to the saturation pressure, which we measure and want to accurately approximate. It's also why there must be only the two areas (yellow and red) and only the three roots.[18]

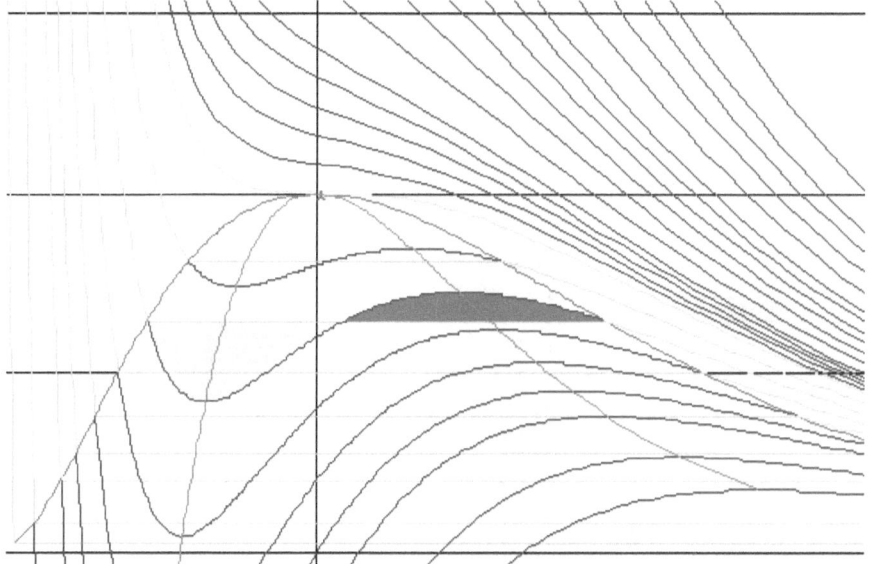

Fugacity Coefficient

It's not enough to simply match *PVT* or even *Z* data. We've introduced the concept of free energy and the importance of accurately representing this should be apparent. Fugacity (given the symbol, *F*) is a sort of pseudo-pressure. It is often defined as the pressure that would result in an ideal gas having the same Gibbs free energy *(g=h-Ts)* as the real gas (at the same temperature). We don't

---

[17] Maxwell's relations are a series of inter-related partial derivatives of the various thermodynamic properties, which are beyond the scope of the current discussion.

[18] Not all empirical equations of state exhibit this behavior, which is problematic and should be avoided, regardless of how well these may seem to represent the data elsewhere.

ever use fugacity in this way, but it does provide a conceptual framework. Most often, we consider the fugacity coefficient (given the symbol, φ), which is the ratio of fugacity to pressure, making it dimensionless. There are several derivations and forms used for the fugacity. The derivations are beyond the scope of this work. For our purposes, we will use the following:

$$\ln \varphi = \int_0^P (Z-1)\frac{dP}{P} \tag{1.18}$$

Equation 1.18 is rarely practical to evaluate. Integration by parts yields the following much more useful formula:

$$\ln \varphi = Z - 1 - \ln Z - \int_\infty^V \left(\frac{P}{RT} - \frac{1}{V}\right) dV \tag{1.19}$$

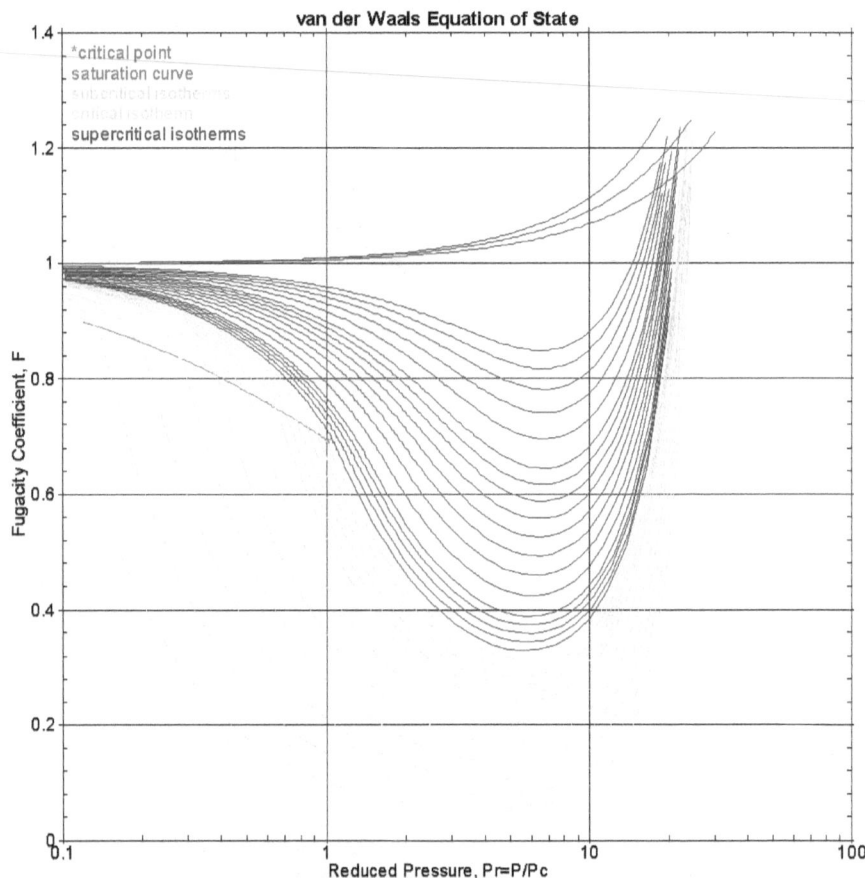

For an ideal gas Z=1 and $\ln\varphi=0$ so that $\varphi=1$ and F=P. In general, the less ideal a fluid behaves, the more Z and φ depart from unity. The preceding figure shows the fugacity coefficient for a van der Waals fluid: Notice that the red saturation curve has collapsed to a single line (i.e., there is no vapor dome). The fugacity of the saturated liquid and vapor is the same, which ties back into the Gibbs free energy and Maxwell's criterion. Notice also that the sub-critical isotherms (green curves) have the same value, but are not continuous in slope, as they cross the saturation curve. We can use the Nelson-Obert data to generate this same curve. You can find all of the digitized data in the same spreadsheet (Nelson-Obert.xls).

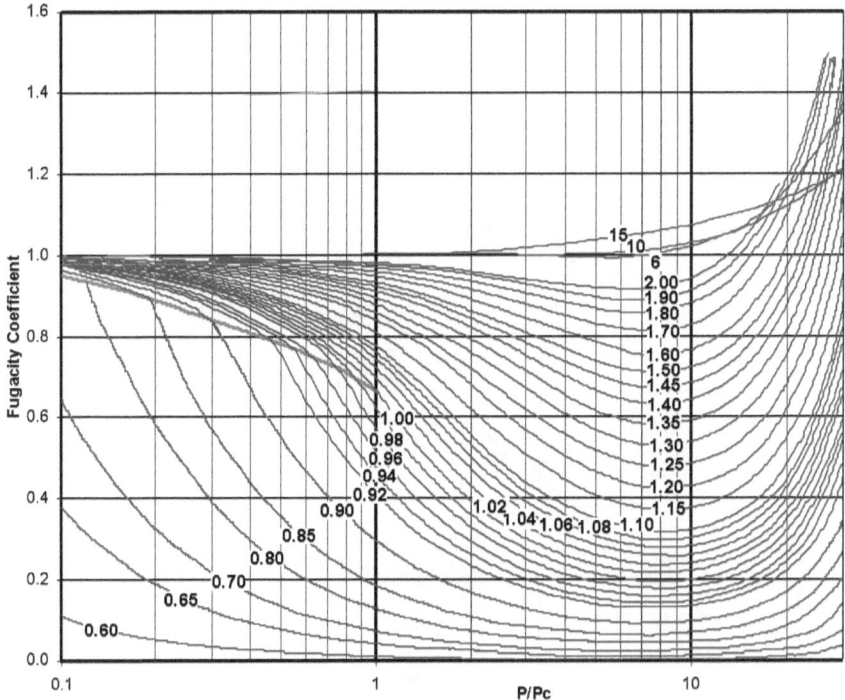

The vertical scale isn't exactly the same as for the van der Waals and the individual isotherms aren't exactly the same, but the general shape is consistent. As noted before with the deviation for sub-critical liquids (at the bottom), we see that this second graph of actual data is deeper and broader, while the super-critical vapors (at the top) are in better agreement. This will not be the case for some of the other simple equations of state.

Residual Enthalpy

We next consider what is called *residual* enthalpy. Like fugacity, this is the part that's not accounted for by ideal gas properties, including temperature-

21

varying specific heat. Residual enthalpy is purely a *PV* effect, not a thermal one, which is captured in the integral of the specific heat. While the pressure equation, similar to 1.18 is often presented in textbooks, the far more useful volume integral is used.

$$\frac{H_0 - H}{RT_C} = T_R(1 - Z) - Z_C \int_{\infty}^{V_R} \left(P_R - T_R \frac{\partial P_R}{\partial T_R} \right) dV_R \qquad (1.20)$$

In this equation, H_0 is the ideal gas enthalpy, H is the real fluid enthalpy, R is the ideal gas constant, and T_C is the critical temperature. The combination is dimensionless. As with fugacity, a graph of residual enthalpy can be constructed from the Nelson-Obert data.

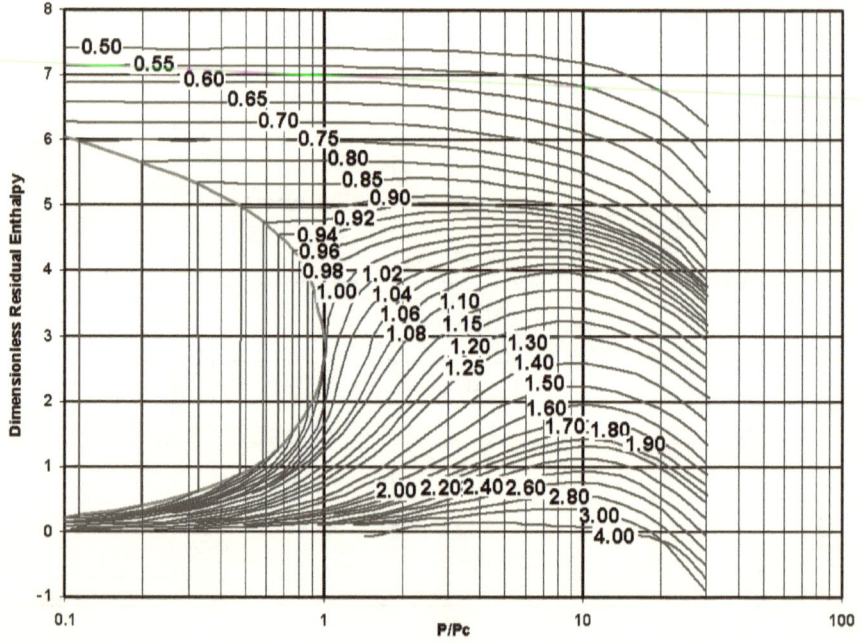

Residual Entropy

We can likewise define a residual entropy as the difference between the ideal gas and actual. There is also a rather useless pressure form of this relationship, along with the volume one listed below:

$$\frac{S_0 - S}{R} = -\ln Z + \int_{\infty}^{V_R} \left(\frac{1}{V_R} - \frac{\partial P_R}{\partial V_R} \right) dV_R \qquad (1.21)$$

Again, S_0 is the ideal gas entropy and S is the real fluid entropy. Through a convoluted process, one can show that the residual entropy is also related to the fugacity coefficient:

$$\frac{S_0 - S}{R} = \frac{H_0 - H}{RT} + \ln F \qquad (1.22)$$

In terms of reduced quantities (more compact notation):

$$S_R = \frac{H_R}{T_R} + \ln F \qquad (1.23)$$

The residual entropy can also be derived from the Nelson-Obert data:

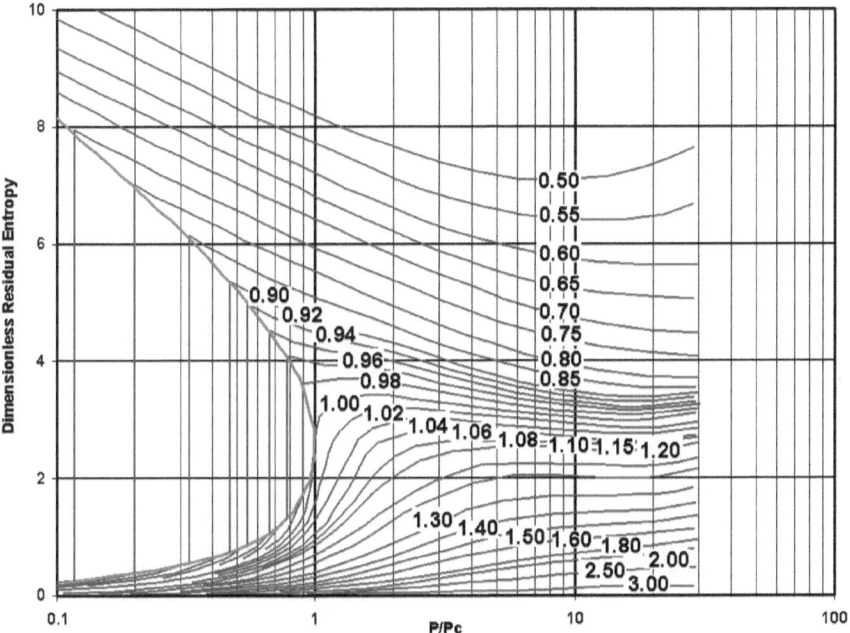

Comprehensive EOS Comparison

We have introduced the PV, PZ, PF, H_R, and S_R charts in order to make a comprehensive equation of state comparison. Some equations of state may do quite well with PV, but not Z or F, while others may do well with H_R and not S_R. In order to accurately approximate real fluid behavior, we must consider all of these. I have carefully digitized all of these curves and you can find them in the Nelson-Obert.xls spreadsheet as well as the csv file in the same folder. With this information, we can now test various equations of state and see how they agree with this extensive data set.[19] All of the various formula can be found in the source code comparison.c. We will not plot every combination of graph with each EOS. I have a provided a program that will so (EOSplt), which you can download free from my web site.

As we will see, using T and V to calculate P, then F, H, and S can result in huge discrepancies, especially for liquids, because dP/dV can be very large. Rather, we use T and P to calculate V, then F, H, and S. This yields much more reasonable results and a more meaningful comparison. We will first consider the Nelson-Obert generalized compressibility chart data compared to the van der Waals EOS.

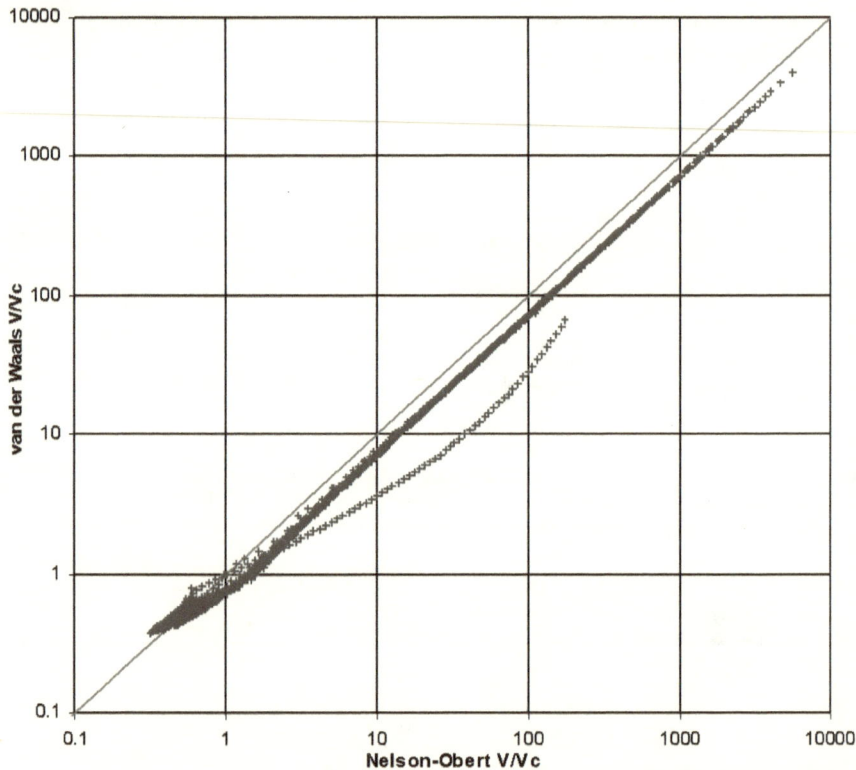

The liquid data (lower left) is close to the exact line and has more scatter than the vapor data (middle and upper right). On average, the calculated values are 22% below the target. This is due to the fact that the Nelson-Obert data have Zc=0.27; whereas, the van der Waals EOS has Zc=0.375. As Vr=ZTr/ZcPr,

---

[19] The Nelson-Obert data only includes simple (well-behaved) fluids. We would expect more difficulty approximating unusual fluids.

most of this offset is due to the 28% discrepancy in the reciprocal of the two critical compressibilities. We see right off that a fixed Zc for an EOS is a drawback. It is preferable that Zc be adjustable.

The arc of points that fall below the diagonal line are the saturated liquid data. The van der Waals EOS is known to be inaccurate for saturated liquids, as the $T/(V\text{-}b)$ term considers the molecules to be hard spheres. The arc will eventually fold back in to the diagonal band of points at increasingly lower values of temperature not shown in the Nelson-Obert data. At very low temperatures, the saturated vapor is essentially an ideal gas.

A comparison of density ($\rho=1/V$) is also revealing. We see that the discrepancies extend over the entire range of densities, not just one portion, and there is no simple pattern or bias. The saturated liquid points are in the bottom left corner and also exhibit an arc shape. The higher density points (compressed liquids) continue upward and only a few points lie on the line.

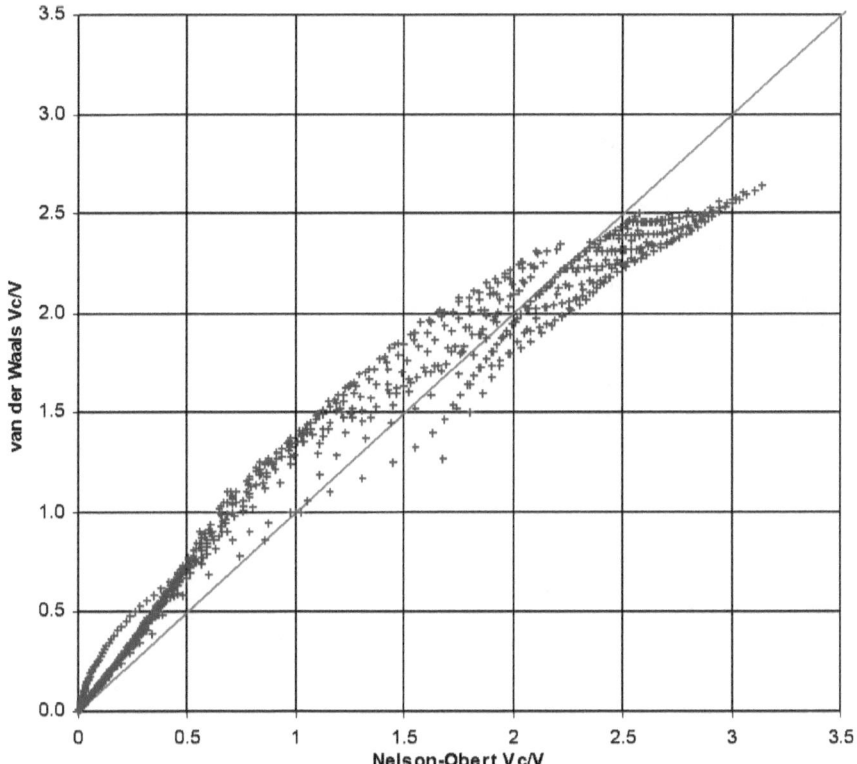

A comparison of pressures is shown in this next figure:

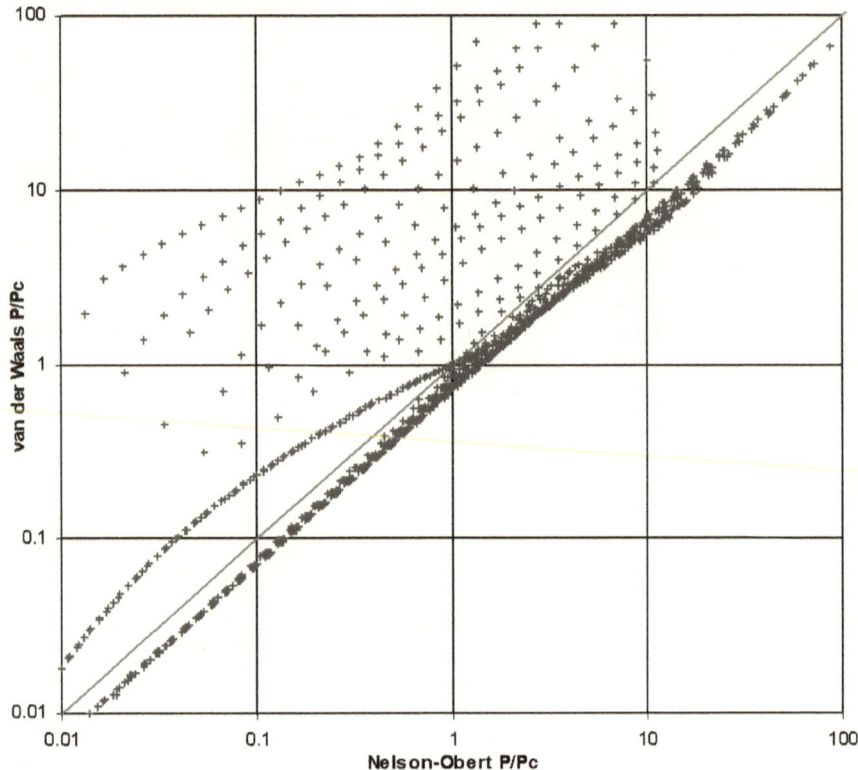

As mentioned previously, the discrepancies in pressure are not as meaningful as those in specific volume or density. Considering this is a log-log plot, some of the differences are more than an order of magnitude.

A comparison of compressibility is shown in this next figure:

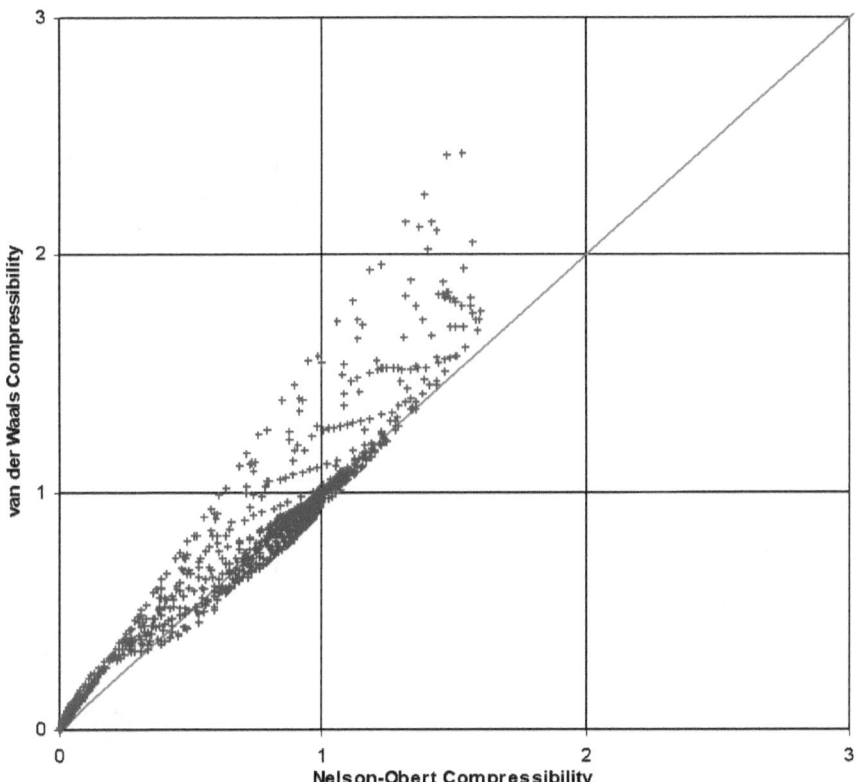

Notice that the cluster of data in the compressibility comparison is similar to the density comparison and not to either the specific volume or pressure. This is typical and why we most often focus on this or the specific volume. Of course, one is a log-log scale and the other is linear. The linear scale is more convenient for accumulating residuals. The same can be obtained with the difference in *log(V)* or *ΔV/V*.

A comparison of fugacities is shown in this next figure:

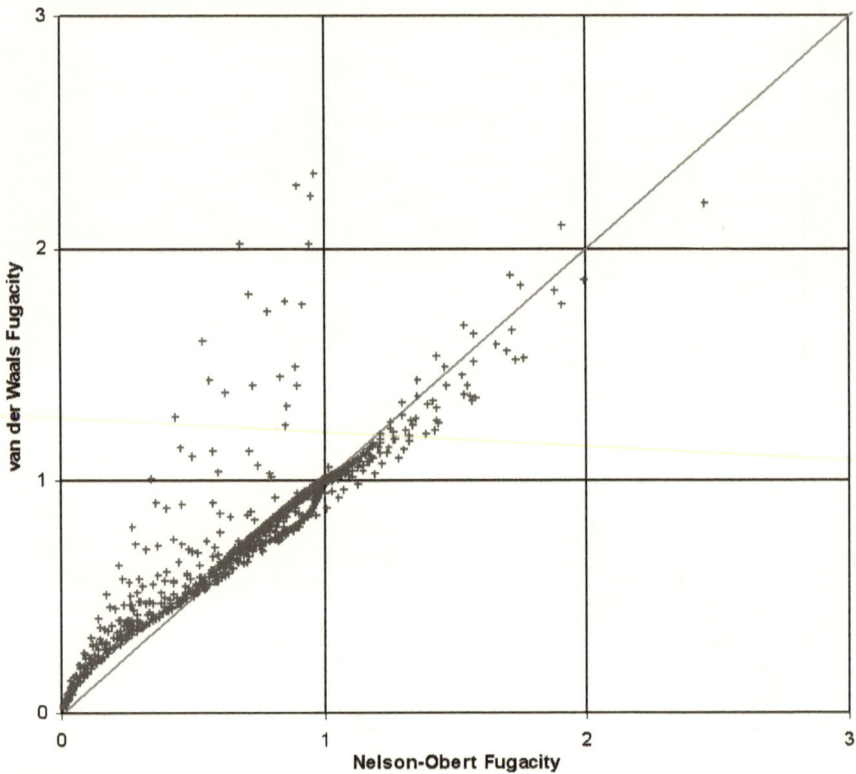

While the discrepancies in these four graphs may not seem too bad so as to completely abandon the van der Waals EOS, the next two will make that clear. This next figure shows a comparison of the residual enthalpies.

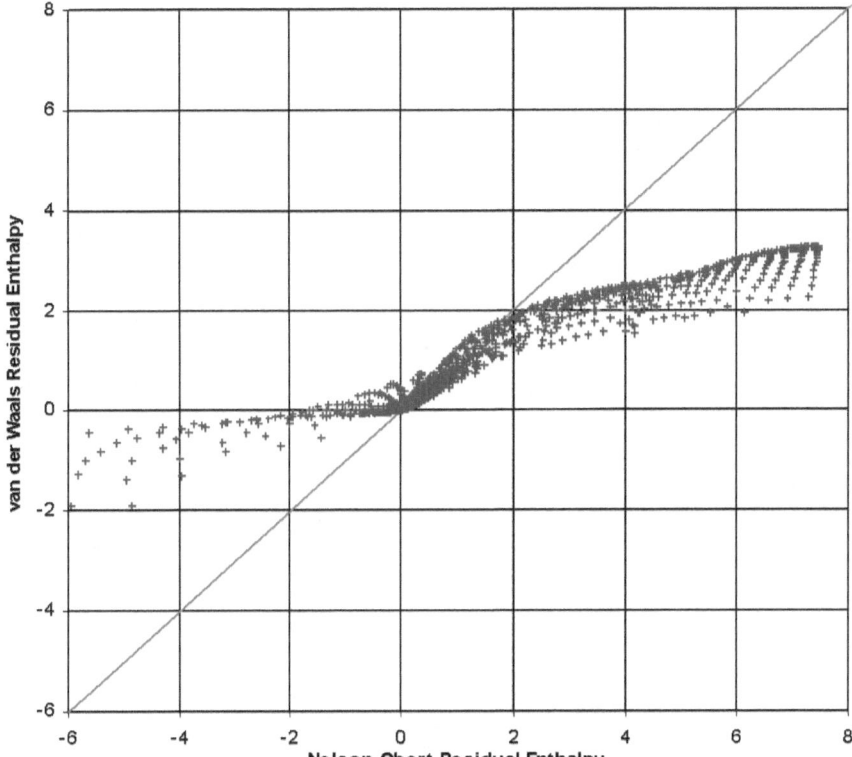

This is really bad. The x and y scales are the same, showing that the behavior is all wrong. We note that the trend is much too flat. This means that the impact of *PV* and *dP/dT* (Equation 1.20) is much too small. We will, therefore, look for equations of state that have more temperature dependence in the various parameters, specifically the term A/V^2. The residual enthalpy accounts for the latent heat of vaporization, a very important property of any fluid. Clearly, the van der Waals EOS doesn't come close to accurately representing this.

This next figure is a comparison of the residual entropies.

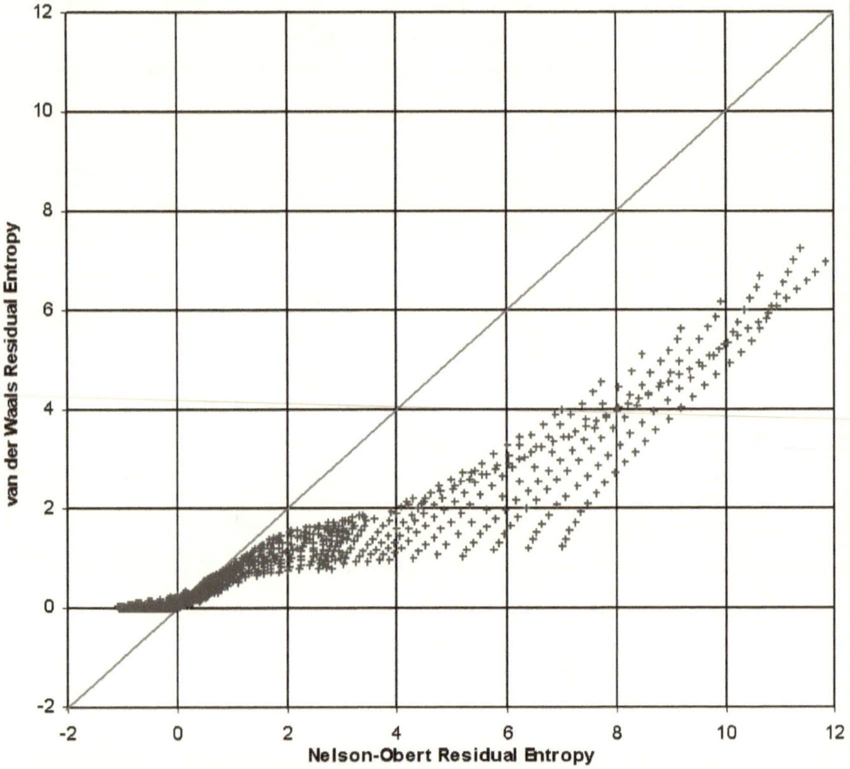

Not only is the shape of the data clustering completely wrong, there is evident bias (on the low side), although the slope is more reasonable than for the residual enthalpy. This is means that the van der Waals EOS more nearly captures dP/dV than dP/dT (compare Equations 1.20 and 1.21).

The program provided (comparison.c) calculates all of the data for all seven figures for each of the equations of state mentioned so far, along with several others that will be presented in the next chapter. The results and graphics can be found in the spreadsheet comparison.xls; therefore, we will only include the most interesting of these graphs here.

The Clausius EOS does not accurately predict densities, as shown in this next figure. This is problem with most simple equations of state and not surprising, because both the pressure and specific volume vary by orders of magnitude. Notice that the densities are consistently high, except for light vapors (bottom left).

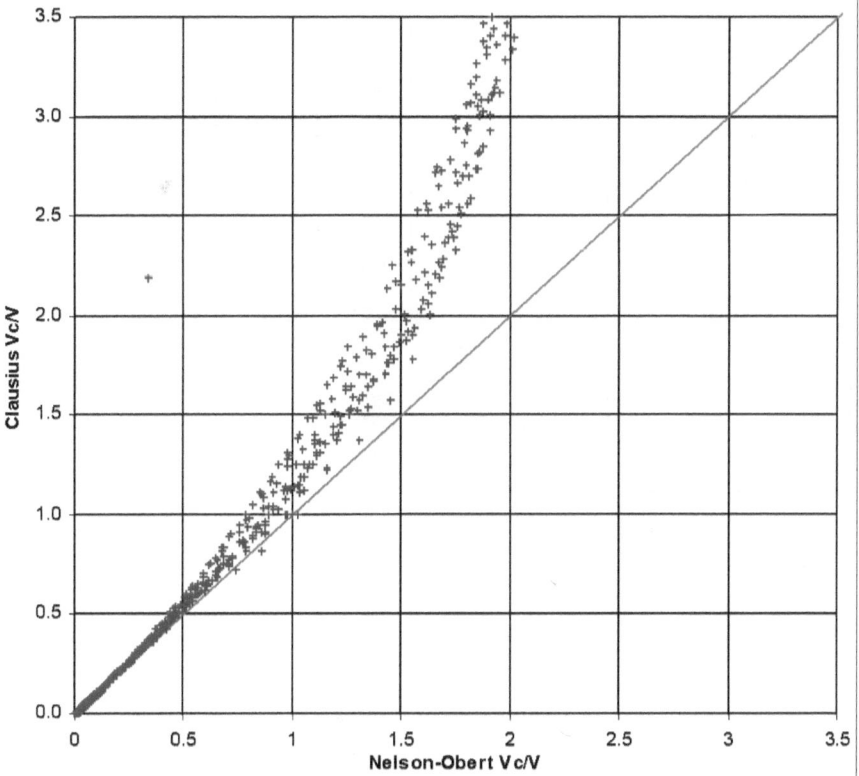

Consistently wrong densities result in consistently wrong compressibilities (as well as fugacities, residual enthalpies, and residual entropies), as shown in this next figure.

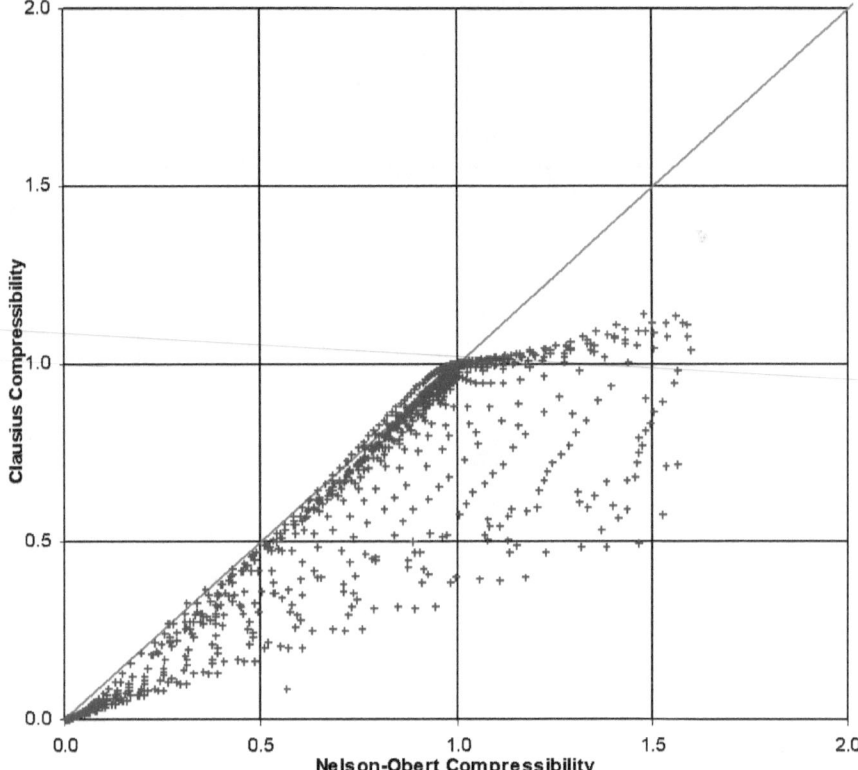

This problem with the Clausius EOS can be significantly improved with a minor modification, which will be covered in the next chapter. The Dieterici EOS is perhaps the least accurate of the seven presented in this chapter, as the following comparison of compressibilities reveals.

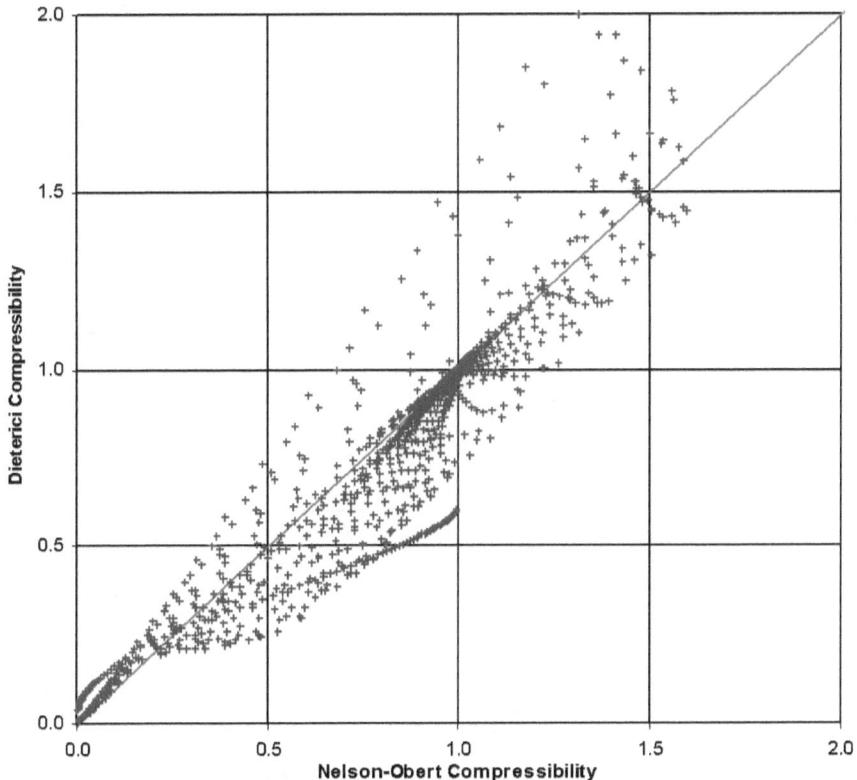

The last comparison we will show in this chapter is for the Redlich-Kwong EOS. As shown in the following figure, this is a considerable improvement over Dieterici.

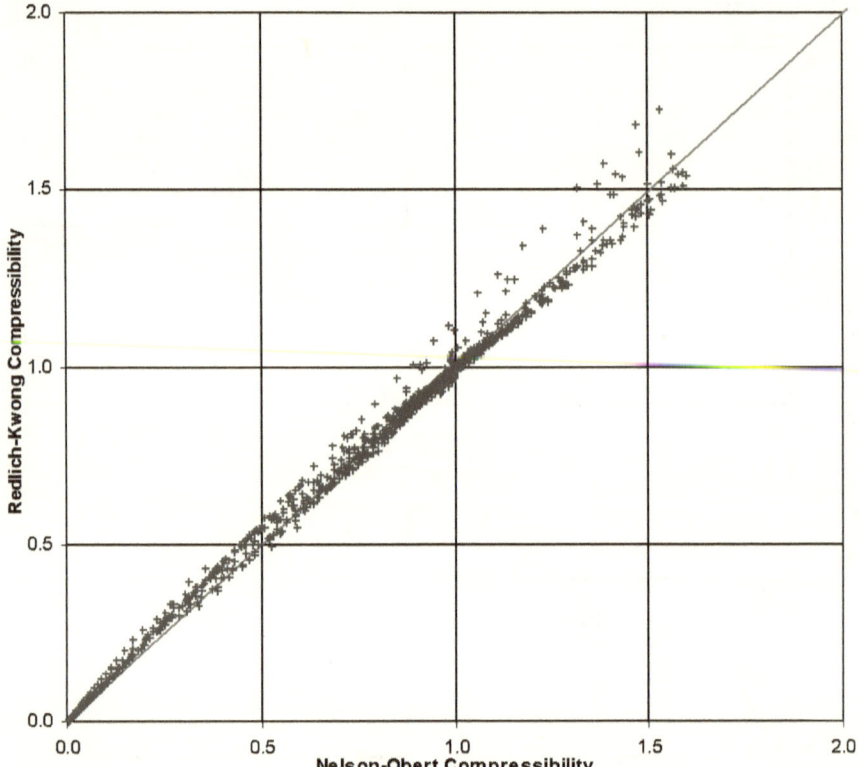

There isn't much point showing any more of these figures. There are 105 of them in comparison.xls. While they're interesting, belaboring the point isn't informative. These do, however, provide motivation for improved equations of state.

Chapter 2. Improved Equations of State

The previous seven simple equations of state are interesting and useful for illustration and academic instruction, but not nearly accurate enough to be of practical use. That's why there are so many *improved* equations of state.

Pitzer Acentric Factor

The Nelson-Obert charts were based on well-behaved fluids (i.e., those having nearly symmetric molecular structure and low bond stresses). Our initial evaluations are necessarily based on this type of fluid, as the N-O data form the basis of our comparisons. There are many fluids that have more complex molecular structures and these have been the motivation for developing enhancements for these seven equations of state. One measure of molecular eccentricity (or asymmetry) is the Pitzer acentric factor, w, defined by the following relationship.[20]

$$w = -\log_{10}\left(P_R^{sat}\right) - 1$$
$$at\, T_R = 0.7$$
(2.1)

This useful parameter is derived from the vapor pressure and is a characteristic of the vapor dome in PV and PZ curves. Noble gases (e.g., argon, krypton, and xenon) have an acentric factor of zero, while water has a value of 0.344 and ethanol 0.644. Note that all of the fluids selected by Nelson and Obert had small (≈ 0) acentric factors. The Pitzer acentric factor is often part of modified equations of state.

Soave's Modification

As we have already presented the Redlich-Kwong EOS, we will first consider Soave's improvement. Soave noted that the van der Waals force (the $-a/V^2$ term) decreased with temperature.[21] Reviewing a variety of fluids led to the following modification of the A term. The rest of the equation is the same as 1.12.

$$A = 0.42748\left[1 + \left(0.480 + 1.574w - 0.176w^2\right)\left(1 - \sqrt{T_R}\right)\right]^2$$
(2.2)

While this modification doesn't have a huge impact on the agreement for specific volume, density, compressibility, fugacity, residual enthalpy, and residual entropy, it's a consistent improvement in all these. The next figure shows the change in compressibility resulting from this modification. Note that

---

[20] Pitzer, K. S., et al., "Volumetric and Thermodynamic Properties of Fluids II: Compressibility Factor, Vapor Pressure, and Entropy of Vaporization," *Journal of the American Chemical Society*, Vol. 77, pp. 3433, 1955.
[21] Soave, G., "Equilibrium Constants from a Modified Redlich-Kwong Equation of State," *Chemical Engineering Science*, Vol. 27, No. 6, pp. 1197–1203, 1972.

the differences are all in the higher densities (liquids), which is where the most improvement was needed.

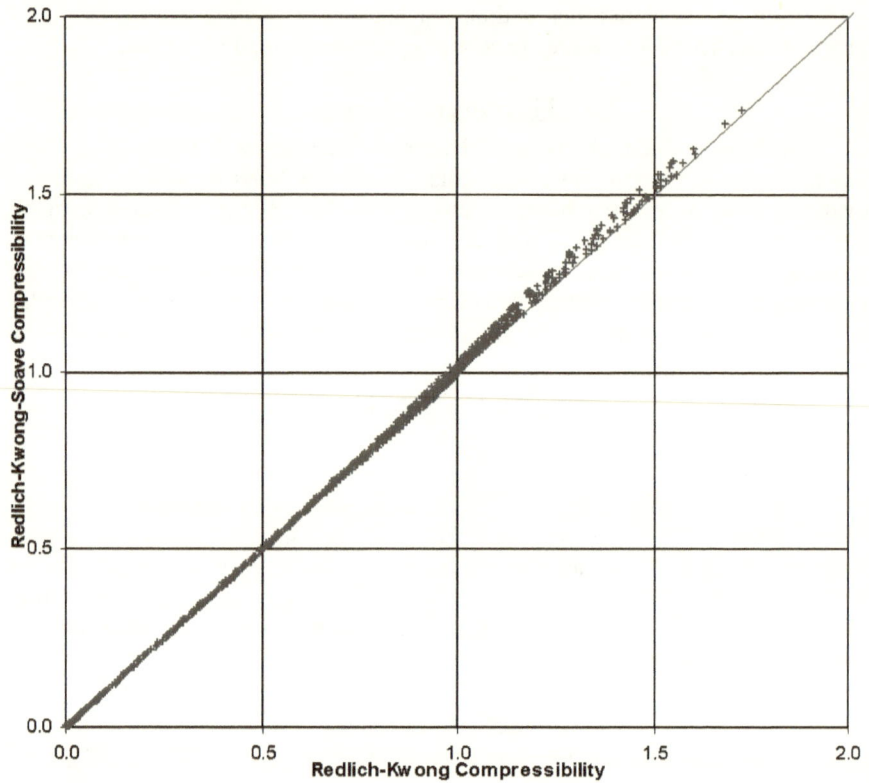

Abbott's Modification

Abbott[22] proposed a similar modification to the Clausius EOS (1.11) by allowing the parameter, A, to vary with temperature and the acentric factor. Abbott's modification is listed below (the rest of the equation is the same:

$$A = \frac{27}{64 Z_C} \left(\frac{1 - \frac{49}{60} w}{T_R^{0.6}} + \frac{49}{60} \frac{w}{T_R^{2.6}} \right) \qquad (2.3)$$

Abbott's modification to the Clausius EOS provides even more improvement in accuracy than the Soave modification to the Redlich-Kwong EOS. A composite of the two compressibility comparisons is shown below:

---

[22] Abbott, M. M., "Cubic Equations of State," AIChE Journal, Vol. 19, p. 596, 1973.

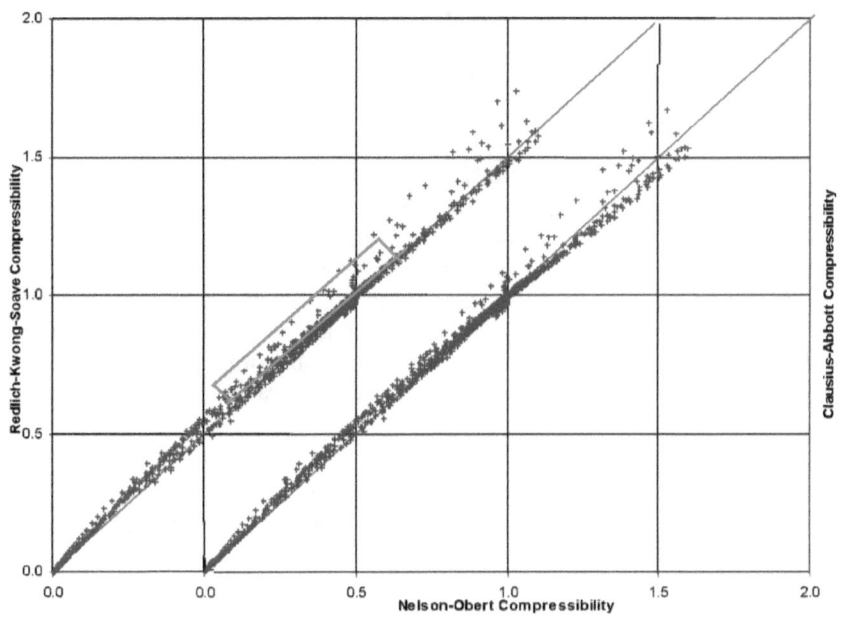

and more noticeable when comparing densities…

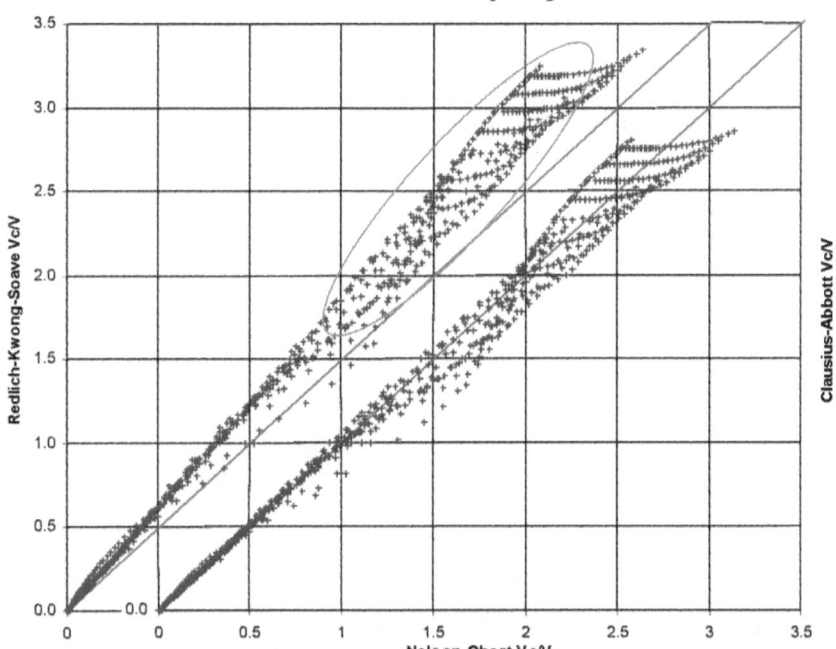

Notice the area of improvement (reduced scatter inside the magenta rectangle) in the middle densities. These two modifications (Soave and Abbott) reveal that the parameter A isn't constant and the numerator in the van der Waals force is also important, that is, $V(V+b)$ vs. $(V+c)^2$.

Modified Boltzmann EOS

After seeing an improvement with the Soave and Abbott modifications to the van der Waals force term, we also consider modifying the Boltzmann EOS (Equation 1.10) by adding $A/\sqrt{T_R}$. Unfortunately, this reduces the accuracy, as you can see on the "modified Boltzmann" tab of the comparison.xls spreadsheet. It's not worth repeating the figures here, but the statistical summary of this modification will be included in the forthcoming table of results. What this does show is that any modifications must be carefully crafted in order to achieve improvements.

Fuller's Modification

Fuller also published a modification to the Redlich-Kwong EOS.[23] In this development, Fuller focused on the liquid (high density) region and the shortcomings of several other cubic equations of state. His article provides an excellent historical overview and is well worth reading. In addition to varying the A parameter with temperature, the denominator is also slightly different.

$$Z_C P_R = \frac{T_R}{(V_R - B)} - \frac{A}{V_R(V_R + C)} \qquad (2.4)$$

Fuller's modification to the parameter, A, is very similar to Soave's:

$$\frac{A}{A_C} = \frac{1 - \dfrac{49}{60}w}{T_R^{0.6}} + \frac{49}{60}\frac{w}{T_R^{2.6}} \qquad (2.5)$$

[23] Fuller, G. G., "A Modified Redlich-Kwong-Soave Equation of State Capable of Representing the Liquid State," Industrial & Engineering Chemistry Fundamentals, Vol. 15, No. 4, pp. 254-257, 1976.

As shown in this next figure, Fuller's modification is somewhat less accurate than either the Soave or Abbott:

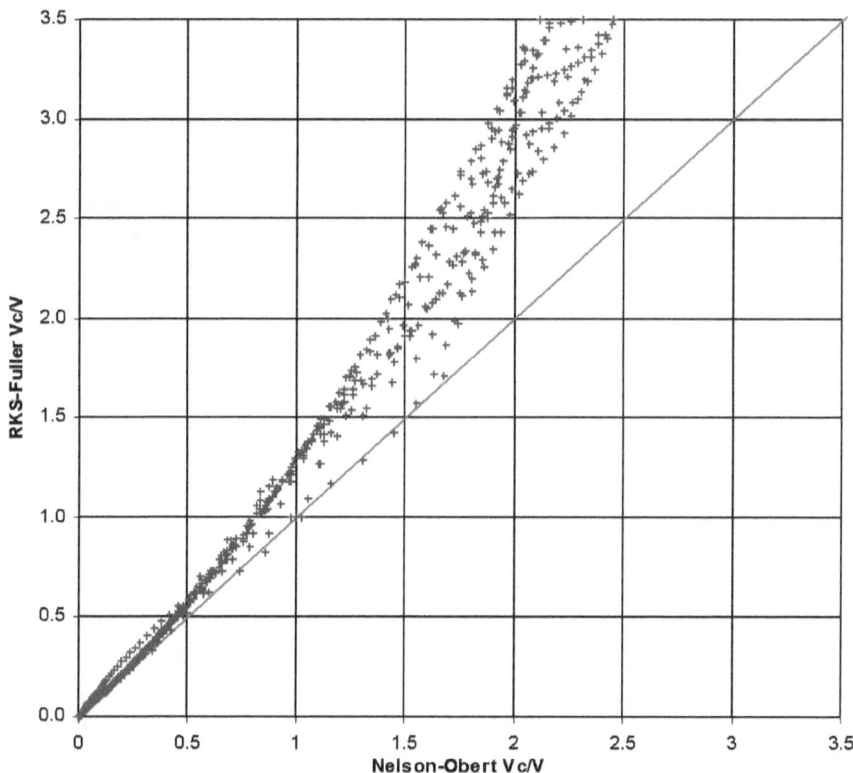

Peng-Robinson EOS

A very successful modification to the basic cubic equation of state has been proposed by Peng and Robinson.[24] This simple change significantly increases the accuracy.

$$Z_C P_R = \frac{T_R}{(V_R - B)} - \frac{A\alpha}{\left(V_R^2 + 2BV_R - B^2\right)} \tag{2.6a}$$

$$A = \frac{0.45724}{Z_C} \tag{2.6b}$$

---

[24] Peng, D. Y. and Robinson, D. B., "A New Two-Constant Equation of State," Industrial and Engineering Chemistry: Fundamentals, Vol. 15, pp. 59–64, 1976.

$$B = \frac{0.0778}{Z_C} \qquad (2.6c)$$

$$\alpha = \left(1 + \kappa\left(1 - Tr^{\frac{1}{2}}\right)\right)^2 \qquad (2.6d)$$

$$\kappa = 0.37464 + 1.54226\omega - 0.26992\omega_2 \qquad (2.6d)$$

where ω is the Pitzer acentric factor. The P-R EOS is often used to calculate the density of natural gas, which is a variable mixture of light hydrocarbons and inert gases. Notice the similarity between the empirical constants in A and B compared to the Redlich-Kwong EOS (Equation 1.12). These too are derived from averages for common fluids.

Author's Modification

These three modifications to the basic cubic equation of state raise the question as to how well could any EOS having this form approximate the data? Instead of varying a single parameter with temperature, why not vary them all? This led the author to consider the following most general form:

$$Z_C P_R = \frac{T_R}{(V_R - B)} - \frac{A}{\left(V_R^2 + CV_R - D^2\right)} \qquad (2.7)$$

You can find the equations, data, and graphics in Benton.xls. This spreadsheet also contains a setup for the Excel® solver to optimize the values of these temperature-dependent parameters so as to minimize the residual in specific volume and compressibility. After applying this process to a dozen different fluids, it became apparent that varying B with temperature is not effective. It also appears that dividing the critical values of A, C, and D by the reduced temperature raised to some power (e.g., $A = Ac/T_R^n$) is adequate. With these four parameters, the critical compressibility can be matched as well as the two partial derivatives of pressure at the critical point. All of the code to implement this is in comparison.c.

Compressibilities are in fair agreement, as shown in the following figure:

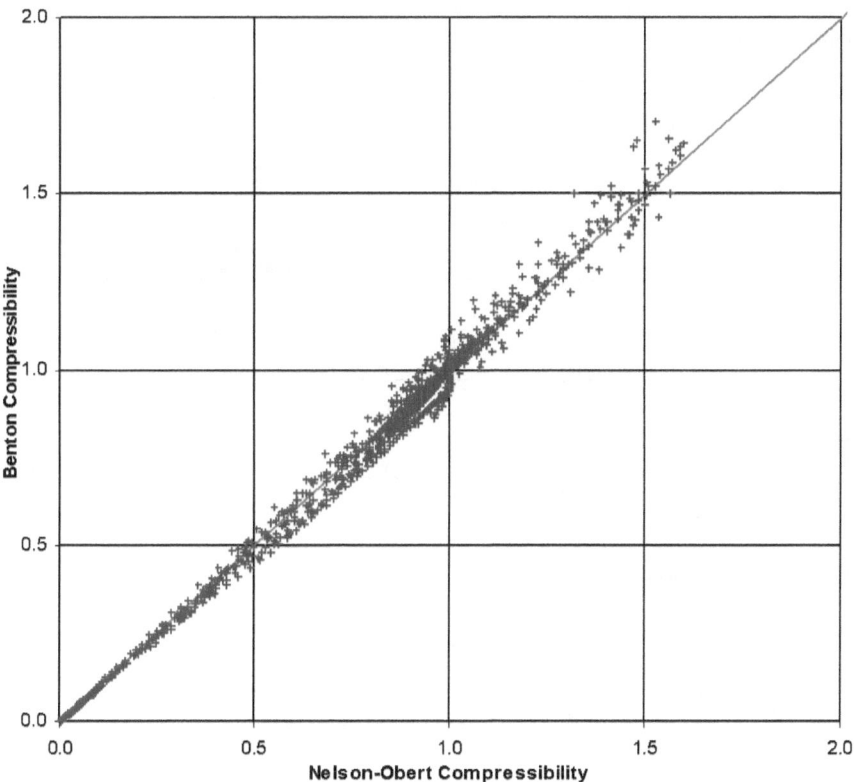

The densities still show bias and artifacts at the high end, indicating that this is as good as it gets with a cubic EOS. Something more is required to achieve better agreement.

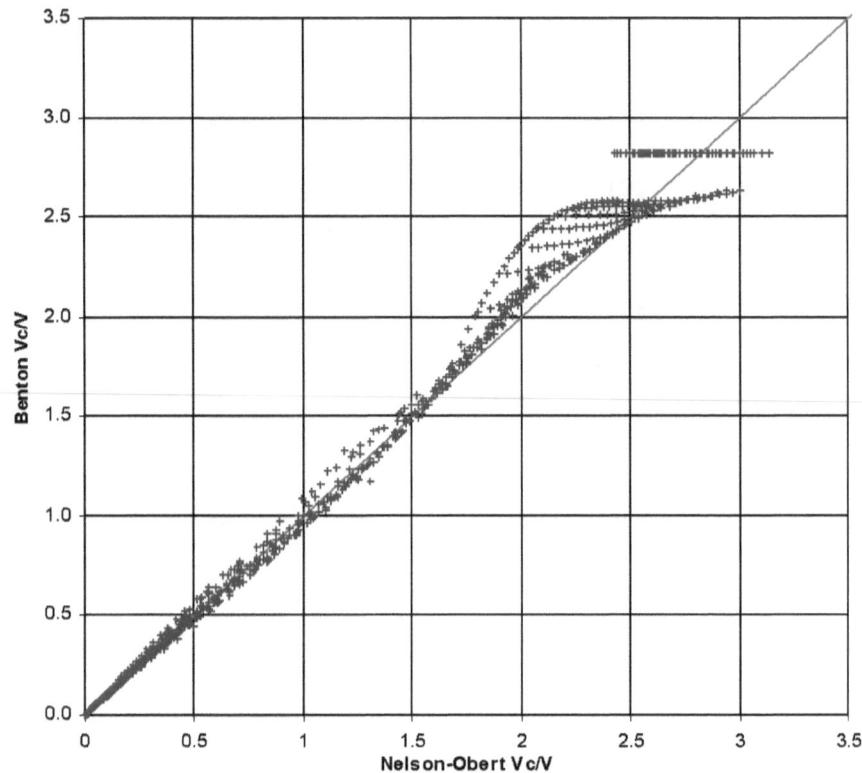

Beattie-Bridgeman EOS

One of the earliest equations of state that depart from the cubic form was developed by Beattie and Bridgeman.[25,26] In reduced form, this EOS can be written:

[25] Beattie, J. A. and Bridgeman, O. C., "A New Equation Of State For Fluids (Part 1)," *Journal Of The American Chemical Society*, Vol. 49, pp. 1665-1667, 1927.

[26] Beattie, J. A. and Bridgeman, O. C., "A New Equation Of State For Fluids (Part 2)," *Journal Of The American Chemical Society*, Vol. 50, pp. 3133-3138, 1928.

$$Z_C P_R = \frac{T_R}{V_R} + \frac{\left(bT_R - a - \dfrac{d}{T_R^2}\right)}{V_R^2}$$

$$+ \frac{\left(-beT_R + ac - \dfrac{bd}{T_R^2}\right)}{V_R^3} + \frac{bed}{V_R^4 T_R^2}$$

(2.8)

The five parameters (a, b, c, d, and e) are adjustable. Two can be controlled to satisfy the critical compressibility and the first derivative of pressure at the critical point; however, if a third is used to satisfy the second derivative of pressure at the critical point, the equation breaks down and the agreement is very poor. You can find this equation along with the data, graphics, and solver setup to find the optimum values for each parameter in Beattii-Bridgeman.xls.

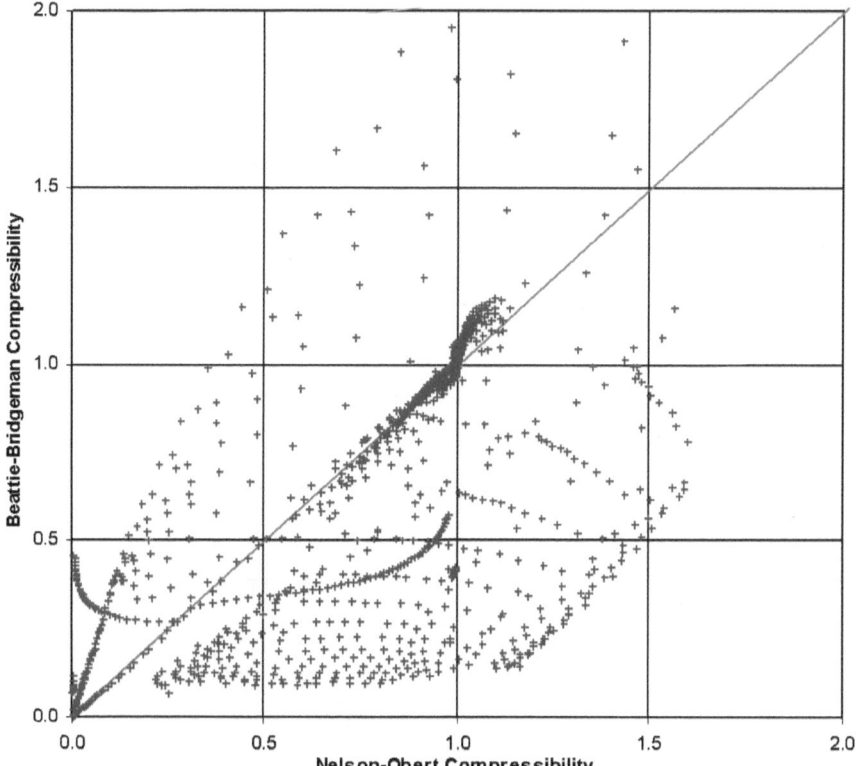

The specific volumes, densities, and compressibilities for this EOS are a mess, as shown by the previous figure as well as the others in comparison.xls. While it is of historical interest, there is little practical use for this EOS.

Benedict-Webb-Rubin EOS

Benedict, Webb, and Rubin also proposed an equation of state similar to Beattie-Bridgeman.[27] Both of these incorporate temperature variation.

$$Z_C P_R = \frac{T_R}{V_R} + \frac{aT_R}{V_R^2} + \frac{cT_R}{V_R^3} - \frac{d}{V_R^2} - \frac{e}{V_R^3} + \frac{f}{V_R^6} - \frac{h}{T_R^2 V_R^2}$$

$$+ \frac{h\left(1 + \dfrac{g}{V_R^2}\right)}{T_R^2 V_R^3 e^{\left(\frac{g}{V_R^2}\right)}} \tag{2.9}$$

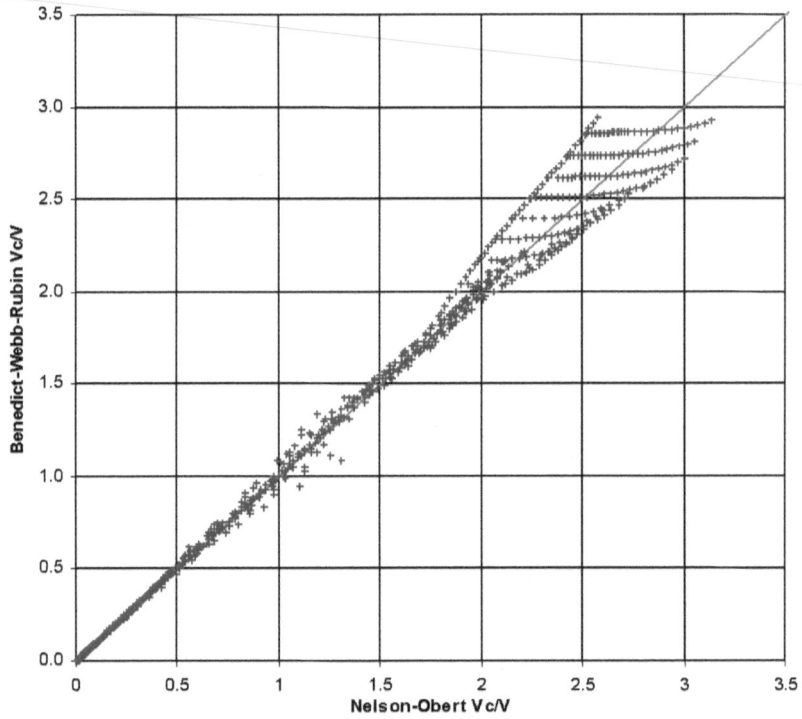

[27] Manson Benedict, M., Webb, G. B. and Rubin, L. C., "An Empirical Equation for Thermodynamic Properties of Light Hydrocarbons and Their Mixtures," *Journal of Chemical Physics*, Vol. 8, pp. 334-345, 1940.

For all of it's appearance through the literature, this EOS doesn't perform any better than the ones already presented in this chapter. The same can be seen in a comparison of compressibilities.

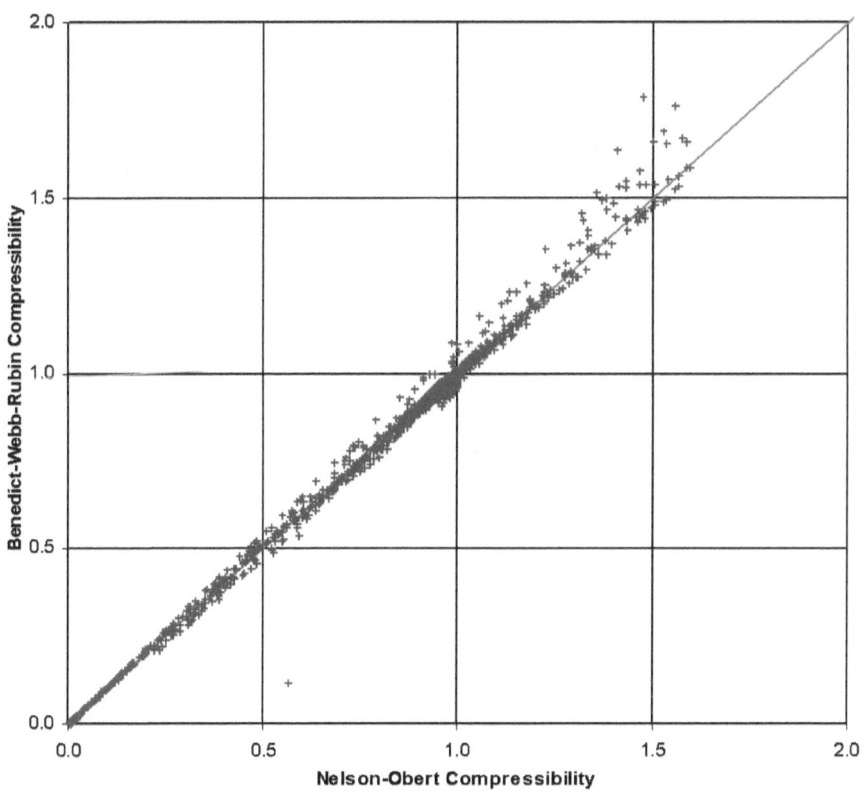

Fifteen EOS Comparison

As mentioned previously, all of the code for the pressure, dP/dV, compressibility, fugacity, residual enthalpy, and residual entropy for all fifteen equations of state presented thus far can be found in comprison.c. The results are written out to comparison.csv, which has been separated into tabs and graphics added in comparison.xls. The summary results are written out to table.csv. The following table is a comparison of all fifteen equations of state for seven different property calculations. The one having the least residual in each category is shown in bold. There is no clear winner and all fall short of the expectations for practical application, which is why we will consider more complicated equations of state, having no simple form.

Table 2.1 Comparison of 15 EOSs

equation of state	root sum squared error						
	V	rho	P	Z	F	H	S
van der Waals	14.6	6.3	40.7	4.4	5.2	86.0	63.4
Beattie-Bridgeman	31.2	34.9	57.3	10.7	621.3	203.0	141.9
Benedict-Webb-Rubin	14.3	2.6	24.3	1.0	**1.2**	197.2	284.2
Benton	**1.6**	**2.1**	17.4	1.1	1.7	**51.0**	36.0
Berthelot	19.5	3.2	32.5	5.0	4.4	66.6	62.6
Boltzmann	11.2	15.5	17.1	1.4	1.7	68.2	27.1
modified Boltzmann	34.5	25.5	47.1	1.8	1.6	112.9	183.5
Clausius	19.8	29.2	26.2	5.3	4.0	86.4	49.5
Clausius-Abbott	7.5	2.7	19.3	0.9	1.7	82.9	75.7
Dieterici	16.1	6.1	18.1	4.0	4.0	79.8	51.4
Peng-Robinson	16.0	7.6	20.3	2.9	3.1	71.4	24.5
Redlich-Kwong	9.7	6.5	17.7	0.9	1.6	71.0	29.0
Redlich-Kwong-Soave	8.8	6.2	**16.3**	**0.9**	1.3	66.8	**21.8**
RKS-Fuller	9.1	13.7	16.5	4.4	3.7	87.2	47.6
Wohl	13.7	21.4	55.0	7.5	7.8	164.0	178.3

Before we move on to steam, there is one more graph we must consider: ZTr vs. Vc/V. I have never seen this graph in any other text. It is interesting because the isotherms are continuous and smooth over the entire range of densities from zero to the solid state. Because the isotherms are smooth and of simple shape, this graph informs us of what isotherms are *supposed* to look like in the meta-stable region under the vapor dome. Notice that the curves never cross. As we will see, this is not the case with the empirical equations of state used for the properties of steam (see corresponding spreadsheets in each of the steam folders).

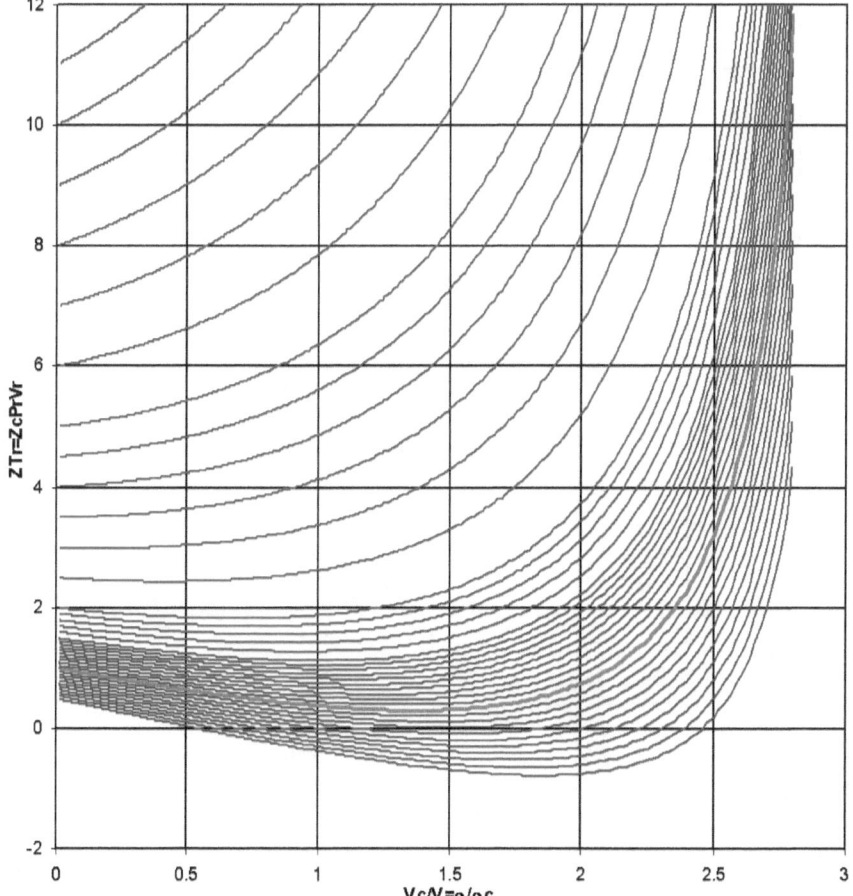

While absent from the literature, this figure is perhaps the most informative representation for a prospective equation of state.

With considerable effort the Nelson-Obert curves can be transformed into the following (Sheet3 in Nelson-Obert.xls macros included):

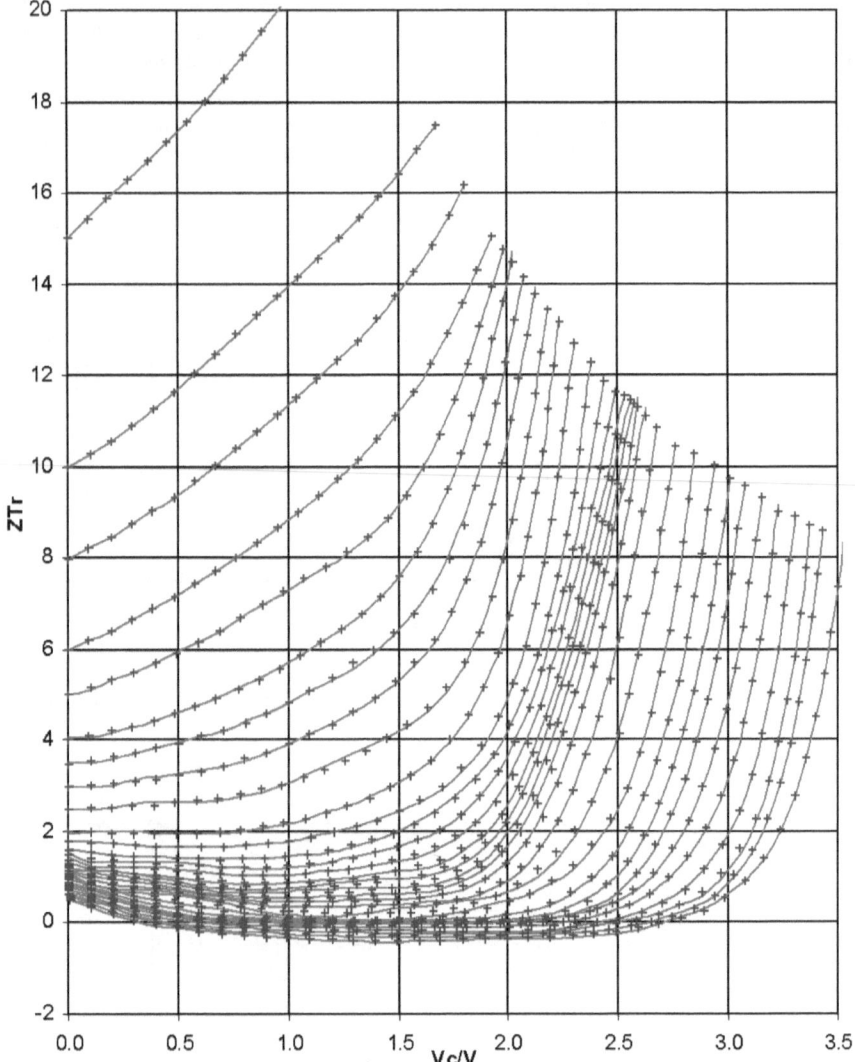

Chapter 3. Steam

Before we discuss any particular empirical equations of state, there is a very important concept that must first be introduced. While I have already provided in the various source codes and spreadsheets in the example\Nelson-Obert folder available free online, this effort was by no means trivial. Calculating the partial derivatives, fixing the criteria at the critical point, and performing the improper (semi-infinite) integral to arrive at the fugacity (necessary for matching the saturation conditions), residual enthalpy, and residual entropy can be daunting. There is no practical closed-form solution for the Dieterici EOS, so that this was accomplished with Gauss quadrature. Integrations are the biggest problem and would greatly limit our choice of form, were these not avoidable.

Rather than starting with an equation for P(T,V), we start with an equation that is already the integral, so that every other property is a partial derivative. This eliminates the need to integrate. The starting point we will use is the Helmholtz free energy $(a=u-Ts)$, as this turns out to be the most convenient. We derive all other properties from this one, through Maxwell's relations already mentioned. For instance, the pressure is found by differentiating with respect to density:

$$p = \rho^2 \frac{\partial a}{\partial \rho} \tag{3.1}$$

You will notice from this point that we often use density (ρ) rather than specific volume (V) as in the first two chapters. Entropy is found by differentiating with respect to temperature:

$$s = -\frac{\partial a}{\partial T} \tag{3.2}$$

Internal energy, u, is easily calculated from a and s:

$$u = a + Ts \tag{3.3}$$

Enthalpy, h, is calculated from internal energy:

$$h = u + \frac{p}{\rho} \tag{3.4}$$

The Gibbs free energy is calculated from the enthalpy:

$$g = h - Ts \tag{3.5}$$

The constant volume specific heat, C_V, is also obtained from the second derivative with respect to temperature:

$$C_V = -T \frac{\partial^2 a}{\partial T^2} \tag{3.6}$$

49

The constant pressure specific, C_P, heat is derived from the constant volume with the density term:

$$C_P = C_V + \frac{T}{\rho^2} \frac{\left(\frac{\partial p}{\partial T}\right)^2}{\frac{\partial p}{\partial \rho}} \tag{3.7}$$

The sound speed, ω, is calculated from these two:

$$\omega^2 = \frac{C_P}{C_V} + \frac{T}{\rho^2}\left(\frac{\partial p}{\partial \rho}\right) \tag{3.8}$$

Once we know the HFE, a, we can readily calculate these other properties. Just as the fugacity, residual enthalpy, and entropy represent the departure from ideal gas behavior, we start with the free energy of the ideal gas. This starting point automatically takes care of the low-density behavior. We then add corrections to account for the real fluid behavior. All of our free energy equations from this point on will have the two parts: ideal plus residual correction. The ideal gas free energy, a_0, has two parts: one arising from the density and a second arising from the specific heat at zero pressure (or density):

$$a_0 = \varphi + RT \ln \rho \tag{3.9}$$

The second term depends on both temperature and density (or specific volume); whereas, the first depends only on temperature. From Equation 3.6 (at zero density), we see that this second term above is in fact:

$$\varphi = \iint \frac{C_{V0}}{T} dT \tag{3.10}$$

The specific heat of most fluids in the rarified gas state (i.e., at vanishing density and pressure) is readily available; therefore, equation 3.10 can be easily determined. There are many and diverse approaches to approximating the residual free energy. We will now consider the three most tried and true ones, which were all developed for water.

Keenan, Keyes, Hill, and Moore EOS

No one has used Keenan, Keyes, Hill, and Moore's (KKHM) 1969 steam properties[28] for decades. They were never widely used, even when they first

---

[28] Keenan, J. H., Keyes, F. G., Hill, P. G., and Moore, J. G., *Steam Tables*, John Wiley & Sons, Inc., 1969.

came out. They were a little more accurate than the 1967 properties[29], which had been endorsed by the ASME and still used by the General Electric Steam Turbine Division to this day (ASME67). There are three more sets of steam properties you should know about: the 1984, published by the NBS/NRC[30] (the National Bureau of Standards is now called the National Institute of Standards and Testing), the IAPWS-SF95[31], and the IAPWS-IF97.[32]

The KKHM properties are based on an elegant formulation and for that reason alone are of significant historical value. The NBS/NRC took this one step farther. It is a great shame that these two works have fallen into disuse. The IAPWS-SF95 is also an elegant formulation, though this work would not have been possible without the trailblazing authors of KKHM and NBS/NRC. The ASME67 and IAPWS-IF97 (based on temperature and pressure, rather than temperature and density) are both mathematical atrocities and should have been tossed in the recycle bin long ago. We will only discuss the first three.

Why are steam properties so computationally intensive? Because we most often know temperature and pressure, but these are not always independent variables and, therefore, can't adequately represent the behavior of a substance over a wide range of conditions. All chemically-stable fluids exhibit saturation states in which the liquid or solid and vapor are in equilibrium. These same substances also exhibit two very interesting points: 1) the critical point, at which the liquid and vapor are indistinguishable, and 2) the triple point, at which the solid, liquid, and vapor coexist in equilibrium.

Temperature and density are always independent for such substances and the properties are continuous in these two variables. Therefore, the formulation must also be continuous in these two variables. This is why the ASME67 and IAPWS-IF97 are mathematical atrocities—they aren't continuous and are jumble of brute-force curve fitting. Keenan, Keyes, Hill, and Moore weren't the first to publish it, but they were the first to implement this important observation: all of the thermodynamic properties of a substance can be expressed in terms of the Helmholtz Free Energy (HFE) or its derivatives. Find the HFE ($a=u-Ts$) and you have the rest!

---

[29] Meyer, C. A., McClintock, R. B., Silvestri, G. J., and Spencer, R. C., Jr., *Thermodynamic and Transport Properties of Steam*, American Society of Mechanical Engineers, 1967.

[30] Haar, L., Gallagher, J. S., and Kell, G. S., *Steam Tables*, NBS/NRC printed by Hemisphere, distributed by McGraw-Hill, 1984.

[31] Wagner, W., and Pruß, A., "The IAPWS Formulation 1995 for the Thermodynamic Properties of Ordinary Water Substance for General and Scientific Use," Journal of Physical Chemistry, Ref. Data 31, pp. 387-535, 2002.

[32] Research and Technology Committee on Water and Steam in Thermal Power Systems, *ASME Steam Properties for Industrial Use*, The American Society of Mechanical Engineers.

KKHM began by splitting the HFE into two parts: one at zero density (i.e., rarified gas) plus a second at finite density that vanishes as the density approaches zero. The first part is:

$$a_0 = \sum_{i=1}^{8} \frac{C_i}{\tau^{i-1}} + C_7 \ln(T) + C_8 \frac{\ln(T)}{\tau} \qquad (3.11)$$

where τ is a dimensionless inverse of the temperature:

$$\tau = \frac{1000}{T} \qquad (3.12)$$

where T is the temperature in °K. The whole expression for HFE, a, is:

$$a = a_0 + RT(\ln \rho + \rho Q) \qquad (3.13)$$

where ρ is the density in gm/cm³, R is the ideal gas constant (0.46151 J/gm/°K), and Q is the *partition* function, which is:

$$Q = (\tau - \tau_c) \left\{ \sum_{j=1}^{7} (\tau - t_j)^{j-2} \left[\sum_{i=1}^{8} A_{i,j} (\rho - r_j)^{j-1} + e^{-E\rho} \sum_{i=9}^{10} A_{i,j} \rho^{i-9} \right] \right\} \qquad (3.14)$$

where $t_j = \tau_c$ for j=1, otherwise $t_j = 2.5$, and rj=0.634 for j=1, otherwise 1. Pressure is given by:

$$p = \rho RT \left(1 + \rho Q + \rho^2 \frac{\partial Q}{\partial \rho} \right) \qquad (3.15)$$

Enthalpy is given by:

$$h = RT \left(\rho \tau \frac{\partial Q}{\partial \tau} + 1 + \rho Q + \rho^2 \frac{\partial Q}{\partial \rho} \right) + \frac{d(\psi_0 \tau)}{d\tau} \qquad (3.16)$$

Entropy is given by:

$$s = -R \left(\ln \rho + \rho Q - \rho \tau \frac{\partial Q}{\partial \tau} \right) - \frac{d\psi_0}{dT} \qquad (3.17)$$

We will first examine the results and then discuss how the process of matching laboratory data was accomplished. Then we will develop software (available free online at my website listed in the forward) to perform this task for any fluid. We begin by graphing the PVT behavior of what we will refer to as KKHM steam.

The reason KKHM used the form of equation 3.14 is that all but the $j=1$ terms disappear at the critical temperature ($\tau = \tau_C$). This allowed them to optimize the parameter selection along the critical isotherm separately from the rest of the laboratory data. This also facilitated the determination of their dimensional factor E. Something that isn't apparent from the original equation form is that subtracting densities $(\rho - \rho_{a,j})$ reduces the magnitude of the coefficients and the round-off error. We will discuss these important details subsequently.

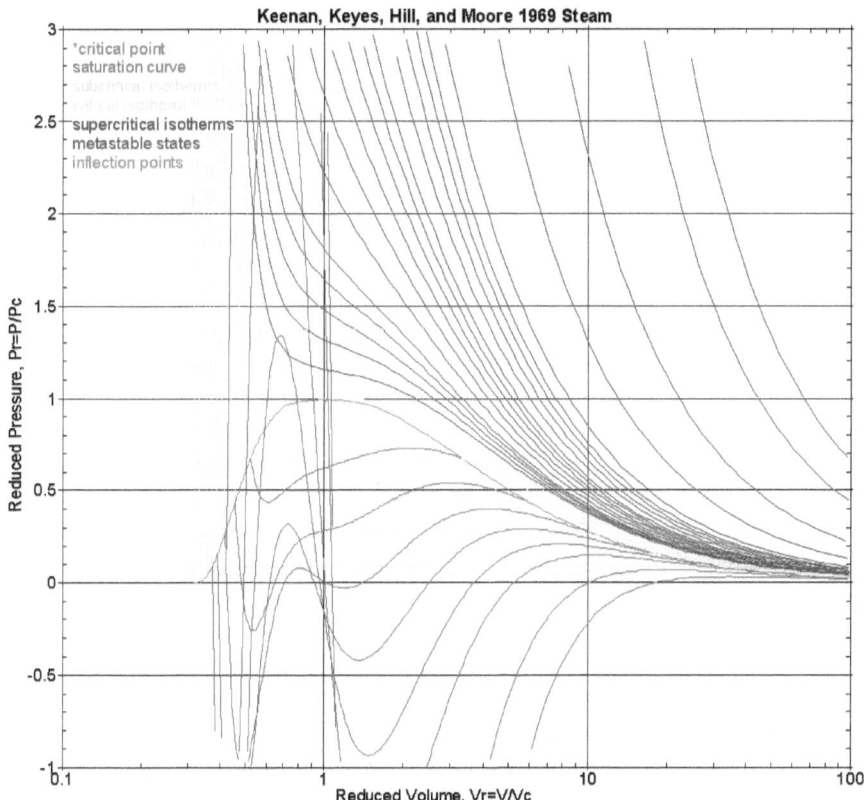

Note the wild behavior of the brown meta-stable isotherm regions below the vapor dome (red curve) and just to the left of the critical isochore ($V_R=1$).[33] Note also that there are more than three roots at temperatures approaching the critical isotherm (cyan curve). KKHM mention this anomaly in their appendix, but do not attribute any significance to it. There is some debate in the literature concerning this all-too-common artifact of empirical equations of state. It is my personal opinion that such should be avoided.

The other figures (compressibility, fugacity, residual enthalpy, and residual entropy) are similar to those already presented for the various simple equations of state for general fluids. We will only include the fugacity here. All of these

---

[33] isothermal=constant temperature, isobaric=constant pressure, isochoric=constant volume, isenthalpic=constant enthalpy, isentropic=constant entropy, … An isochore is a vertical line on a PV plot; whereas, an isobar is a horizontal one on this same figure.

can be displayed with EOSplt, which is available free on my web site listed in the forward. C source code for the KKHM steam properties as well as an Excel® Add-In may also be found on the web site.

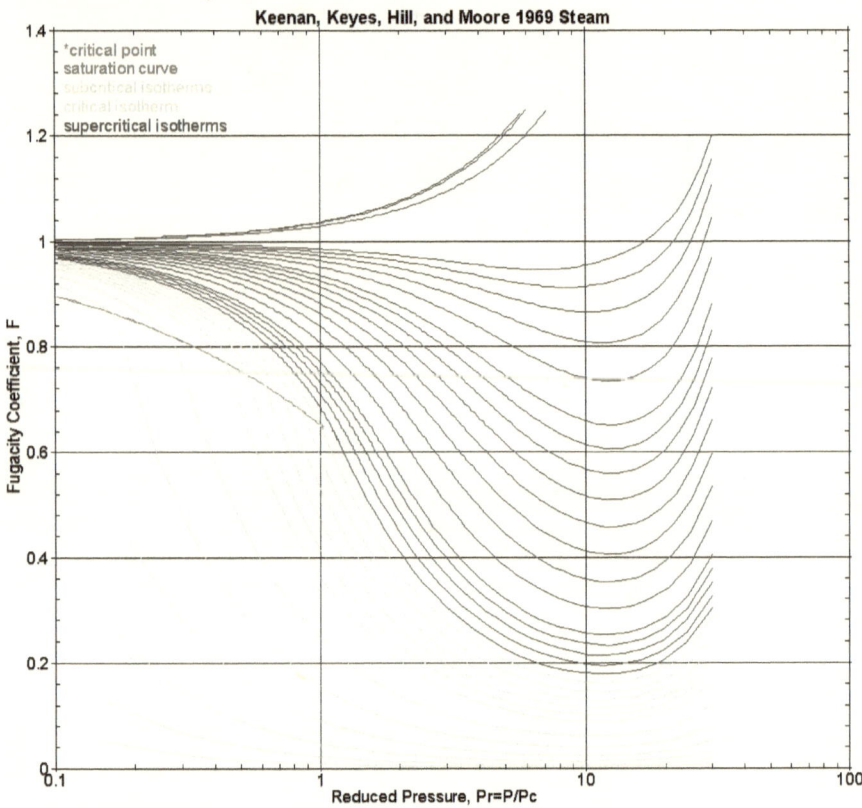

Keenan, Keyes, Hill, and Moore 1969 Steam

The KKHM EOS has been implemented (KKHM.c) and may be found in the examples\KKHM folder. A snippet of this code is listed below:

```
double fA(double T) /* zero density Helmholtz free
    energy a=u-Ts */
    {
    int i;
    double a,tau,tt;
    tau=1000./T;
    tt=1.;
    for(a=i=0;i<6;i++)
        {
        a+=C[i]/tt;
        tt*=tau;
        }
    a+=(C[6]+C[7]/tau)*log(T);
    return(a);
```

```
  }
double fQ(double tau,double rho)
  {
  int i,j;
  double dt,q,r,rhoa,s,tauc;
  tauc=1000./Tc;
  dt=1.;
  rhoa=0.634;
  for(q=j=0;j<7;j++)
    {
    r=1.;
    for(s=i=0;i<8;i++)
      {
      if(j>1&&i>3)
        break;
      s+=A[7*i+j]*r;
      r*=rho-rhoa;
      }
    s+=exp(-E*rho)*(A[7*8+j]+A[7*9+j]*rho);
    q+=dt*s;
    if(j==0)
      dt=tau-tauc;
    else
      dt*=tau-2.5;
    rhoa=1.;
    }
  return(q);
  }
double fP(double T,double rho)
  {
  double tau;
  tau=1000./T;
  return((((fQr(tau,rho)*rho+fQ(tau,rho))*rho
    +1.)*rho*R*T);
  }
double fH(double T,double rho)
  {
  double a,at,Q,Qr,Qt,tau;
  tau=1000./T;
  Q=fQ(tau,rho);
  Qr=fQr(tau,rho);
  Qt=fQt(tau,rho);
  a=fA(T);
  at=fAt(T);
  return(R*T*(rho*tau*Qt+1.+rho*Q+rho*rho*Qr)+a+tau*at);
  }
double fS(double T,double rho)
  {
  double at,Q,Qt,tau;
```

```
tau=1000./T;
Q=fQ(tau,rho);
Qt=fQt(tau,rho);
at=fAt(T);
return(-R*(log(rho)+rho*(Q-tau*Qt))+at*tau/T);
}
```

A sample of the output is listed below:

```
1969 Steam Properties
developed by Keenan, Keyes, Hill, and Moore
implemented by Dudley J. Benton
    Ts          Ps         Vf         Vg            Hf       Hg
     Sf         Sg
273.160   0.000612049 1.00024 205894.83486      0.0  2501.4
   -0.0001 9.1557
275.000   0.000698883 1.00016 181519.73355      7.8  2504.8
    0.0281 9.1069
645.000  21.506997445 2.35504        4.39484  1932.4 2275.0
    4.1738 4.7051
647.245  22.089400000 3.15637        3.15637  2099.5 2099.5
    4.4299 4.4299
    T      P          V          H        S
   275   0.01        1.00015      7.8   0.0281
  1600   0.01    73841.69017   5480.8  11.6259
  1600   0.02    36920.81822   5480.8  11.3060
  1600   0.04    18460.44836   5480.8  10.9861
  1600 163.84        4.87761   5195.0   6.9474
```

Linear Least-Squares with Linear Constraints

Before we can implement equations 3.11 through 3.17, we must discuss how to fit data (i.e., determine coefficients that minimize some residual) and also meet one or more specific objectives (i.e., three conditions at the critical point). Minimizing a residual can be cast in terms of linear least-squares if we're careful to form our equations as a linear combination of terms. This process can be described by the following matrices:

$$X = \left[A^T A\right]^{-1} [A^T B] \qquad (3.18)$$

where A is a rectangular matrix (far more rows than columns) containing the terms evaluated at each data point, B is a column matrix (one column, same number of rows as A) containing the data values, and X is the column matrix containing the coefficients (one column, same number of rows as A has columns). The T indicates the transpose and $^{-1}$ indicates the inverse.

The constraints must also be linear unless we intent to use a nonlinear method, such as the Excel® solver, which is used in the three spreadsheets (Beattit-Bridgeman.xls, Benedict-Webb-Rubin.xls, and Benton.xls). You can find code to perform this nonlinear minimization on my web site, but it's not

56

necessary for our purposes here. The linear constraints can be described by the following matrices:

$$CX = D \qquad (3.19)$$

where C is a rectangular matrix representing the terms and D is a column matrix representing the objectives. Combining these two is done with Lagrange multipliers. The resulting augmented matrix defines the complete problem.

$$\begin{bmatrix} A & C^T \\ C & 0 \end{bmatrix} \begin{bmatrix} X \\ \Lambda \end{bmatrix} = \begin{bmatrix} B \\ D \end{bmatrix} \qquad (3.20)$$

where Λ is a column vector containing the Lagrange multipliers. In some problems the multiplier values are of interest, but this is not the case here.

Implementation

Single-precision floating-point calculations wouldn't get us close to acceptable coefficients and even double-precision is inadequate for this task. KKHM carried out this process on an IBM® mainframe computer in FORTRAN, which has a native quadruple-precision floating-point type. While Intel® floating-point processing units (FPU) perform operations in 80 binary bits—even better than double-precision—they don't achieve anything near quadruple precision.[34]

Some graphical processing units (GPU) provide quadruple precision, but I don't write code that will only run on one very particular super special type of ridiculously expensive hardware; nor do I write code that will only run on some arcane operating system. Therefore, I have provided the code to perform this task in C++, along with the necessary implementation to achieve whatever level of precision you have patience to await the results. While this is neither arbitrary nor infinite precision, it is expandable. You just set the desired number of bytes in xreal.cpp before compiling. The difference between my implementation (which I originally wrote in assembler in 1983) and others is that the binary storage is identical to 8-byte (64-bit) IEEE double-precision, only with more or less bytes in the mantissa. We will discuss this in a subsequent section.

In the folder examples\KKHM you can find a program (MAKEOS.c) that performs the necessary steps to implement linear least-squares with linear constraints and obtain an accurate and continuous correlation for the Helmholtz free energy. The general process can be seen in the main program:

---

[34] Note: Earlier versions of the Microsoft® C compiler (e.g., 7.2) implement 80-bit floating point. Later versions do not. Recent versions of Microsoft® Visual Studio® ignore the long double type and simply replace this with simple double (i.e., 64-bit). I still have a copy of MSC7.20 if you really need it for something special, but it will only produce 16-bit executables, which will not run on a 64-bit version of Windows®.

```
int main(int argc,char**argv,char**envp)
  {
  ReadTitle();
  LinearLeastSquaresLinearConstraints(C,0.,
    0.,Na,Nc,INITIALIZE);
  ReadCriticalData();
  ReadSpecificHeatData();
  ReadSaturationData();
  ReadSuperheatData();
  LinearLeastSquaresLinearConstraints(C,0.,
    0.,Na,Nc,SOLVE);
  ListCoefficients();
  ListCriticalParameters();
  SpecificHeatAgreement();
  SaturatedAgreement();
  SuperheatedAgreement();
  return(0);
  }
```

Constraints at the critical point are added in this section of code:

```
void ReadCriticalData()
  {
  /* first constraint: Z(Tc,Vc)=Zc */
  setdouble(X,Nf,0.);
  Qsub00(Tau,Rho,Q00);
  Qsub01(Tau,Rho,Q01);
  for(i=0;i<Nq;i++)
    X[Nf+i]=Rho*(Q00[i]+Rho*Q01[i]);
  Y=Zc-1.;
  LinearLeastSquaresLinearConstraints(X,Y,
    0.,Na,Nc,CONSTRAIN);
  /* second constraint (dP/dV=0) */
  Qsub02(Tau,Rho,Q02);
  for(i=0;i<Nq;i++)
    X[Nf+i]=((Q02[i]*Rho+4.*Q01[i])*Rho+2.*Q00[i])*Rho;
  Y=-1.;
  LinearLeastSquaresLinearConstraints(X,Y,
    0.,Na,Nc,CONSTRAIN);
  /* third constraint (dýP/dVý=0) */
  Qsub03(Tau,Rho,Q03);
  for(i=0;i<Nq;i++)
    X[Nf+i]=((Q03[i]*Rho+7.*Q02[i])*Rho
    +10.*Q01[i])*Rho+2.*Q00[i];
  Y=0.;
  LinearLeastSquaresLinearConstraints(X,Y,
    0.,Na,Nc,CONSTRAIN);
```

Expansions of Q(tau,rho) and F(tau) are provided in functions such as:

```
void Fsub0(double Tau,double*F)/* user-defined zero-
    density Helmholtz Free Energy */
```

```
{
int i;
for(i=0;i<6;i++)
  F[i]=powi(Tau,-i);
F[i++]=log(1000./Tau);
F[i]=log(1000./Tau)/Tau;
}
void Qsub00(double Tau,double Rho,double*Q) /* user-
  defined partition function */
{
int i,j,k;
double Rhoj,ttt;
for(k=j=0;j<7;j++)
  {
  if(j==0)
    {
    Rhoj=0.634;
    ttt=1.;
    }
  else
    {
    Rhoj=1.;
    ttt=(Tau-1.544912)*powi(Tau-2.5,j-1);
    }
  for(i=0;i<8;i++)
    if(!(j>1&&i>3))
      Q[k++]=ttt*powi(Rho-Rhoj,i);
  Q[k++]=ttt*expl(-E*Rho);
  Q[k++]=ttt*expl(-E*Rho)*Rho;
  }
}
```

Expansions of the partial derivatives of Q and F are provided in similar functions. The least-squares data are accumulated as follows:

```
void ReadSpecificHeatData()
  {
  n=0;
  while(ReadLine(1))
    {
    if(sscanf(bufr,"%lf,%lf",&T,&Cvo)!=2)
      break;
    Tau=1000./T;
    Fsub1(Tau,F1);
    Fsub2(Tau,F2);
    for(i=0;i<Nf;i++)
      X[i]=-(2.*F1[i]+Tau*F2[i])*Tau*Tau/1000.;
    Y=Cvo;
    LinearLeastSquaresLinearConstraints(X,Y,
    1./Y,Na,Nc,ACCUMULATE);
```

```
    n++;
    }
  }
```

The global arrays and their dimensions are defined as follows:

```
#define Nf  8 /* number of terms in F (Psio) */
#define Nd  1 /* number of density constraints (1 or 2
    if Water; otherwise 0) */
#define Nc  6 /* number of constraints */
#define Nq 50 /* number of terms in Q */
#define Na (Nf+Nq) /* total number of coefficients */
double C[Na];
```

The input data for steam is in **KKHM.csv**, an excerpt appears below:

```
THERMODYNAMIC PROPERTIES OF STEAM (KKHM 1969)
CRITICAL POINT PROPERTIES
Tc,Vc,Pc,MW
647.2861111,3.154572247,22.08935447,18.016
TRIPLE POINT PROPERTIES
To,Vo,Po
273.16,1.000220784,0.000611289
POINT OF MAXIMUM DENSITY DATA
Tx,Vx,Px
277.15,1.000002287,0.250003899
CONSTANT VOLUME SPECIFIC HEAT OF STEAM AT ZERO DENSITY
T,Cvo
277.5944444,1.397595708
283.15,1.39860054
...
1588.705556,2.18844036
SATURATION TVP,DATA
Tsat,Psat,Vf,Vg
274.8166667,0.000700866,1.000152114,180885.0158
277.5944444,0.000852475,1.000114657,150207.916
...
647.2611111,22.0814255,2.930805465,3.411944
SUBCOOLED AND SUPERHEATED TVP DATA
T,P,V
310.9277778,0.006894757,1.006981733
338.7055556,0.006894757,22634.63052
...
1588.705556,137.8951459,5.629890827
```

The output is written to **KKHM.out** (name formed from the input file):

```
MAKEOS/V1.20: Empirical EOS Based on the Helmholtz Free
    Energy
Using Linear Least-Squares Curve-Fit with Linear
    Constraints
```

The Helmholtz Free Energy, A, is defined by the
   equations:
  A=U-T*S   Tr=T/Tc   Vr=V/Vc
   A=R*Tc*(F+Tr*(Q/Vr-ln(Vr)))
      F=F1(Tr)+F2(Tr)+F3(Tr)+...
      Q=Q1(Tr,Vr)+Q2(Tr,Vr)+Q3(Tr,Vr)+...

Number of Terms in F 8
Number of Terms in Q 50
Number of Terms in A 58
Number of Constraints 6
Number of Free Constants 52
Number of Simultaneous Equations 64

Title: THERMODYNAMIC PROPERTIES OF STEAM (KKHM 1969)

Reading Critical Point Data and Molecular Weight
   Tc=647.286 Vc=3.15457 Pc=22.0894
   MW=18.016 R=0.461504 Zc=0.233266
   Adding 3 Constraints at the Critical Point
Reading Triple Point Data
   To=273.16 Vo=1.00022 Po=0.000611289
   Adding 2 Constraints at the Triple Point
Reading Point of Maximum Liquid Density (Water Only)
   Tx=277.15 Vx=1 Px=0.250004
   Adding Constraint at the Point of Maximum Liquid
   Density
Accumulating Least-Squares on Specific Heat Data
   237 data points accumulated
   277.5946T61588.71
Accumulating Least-Squares on Saturation Data
   160 data points accumulated
   274.8176T6647.261
Accumulating Least-Squares on Subcooled and Superheated
   Data
   1505 data points accumulated
   310.9286T61588.71 0.006894766P6137.895
Solving for Coefficients

Results of Constrained Least-Squares
   C[0]=1850.770895632598
   C[1]=3244.355462260212
   ...
   C[57]=68.66569180299729

Comparison with Specified Properties
   critical pnt.: Tc=647.286 Vc=3.15457 Pc=22.0894
                  Hc=2087.86 Sc=4.40763 Ac=-834.82
   triple point: To=273.16 Vo=1.00022 Po=0.00878758

61

```
                    Ho=0.01   So=-0.00000   Ao=-0.00
     max.  density: Tx=277.15  Vx=1  Px=0.221598
                    Hx=19.65   Sx=.f   Ax=0.07


 Constant Volume Specific Heat at Zero-Density
    T   Meas.  Calc.  Error
   278 1.3976 1.3976 -0.0000
   283 1.3986 1.3986 -0.0000
 ...
  1589 2.1884 2.1885  0.0000
              average=-0.0000
              maximum= 0.0000
              std.dev.= 0.0000


 Agreement with Saturation Data
         <-------- Zg -------> <-- Psat*Vfg --->
   Tsat  Meas.  Calc.  Error   Meas  Calc Error
   274.8 0.9996 0.9996  0.0000 126.8 126.7  -0.1
   277.6 0.9995 0.9995  0.0000 128.0 128.0  -0.0
 ...
   647.3 0.2522 0.2522 -0.0000  10.6  10.6   0.0
              average=-0.0003      average= 0.0
              maximum=-0.0018      maximum=-0.1
              std.dev.= 0.0005  std.dev.= 0.0


 Agreement with Superheated and Subcooled Data
              <-------- Z -------->
    T     P      Meas.  Calc.  Error
    311   0.007 0.0000 0.0000 -0.0000
    339   0.007 0.9984 0.9984  0.0000
 ...
  1589 137.895 1.0588 1.0600  0.0011
                   average= 0.0000
                   maximum= 0.0049
                   std.dev.= 0.0006


 Zero-Density Helmholtz Free Energy & Second Virial
    Coefficient
    T       Ao        B
   250   1493.99   -29.22
   300   1318.86   -17.73
 ...
  1600  -5564.92     0.05
 Boyle temperature=1493
```

The resulting coefficients are close, but not exact, to the original KKHM values, as shown in the following figure:

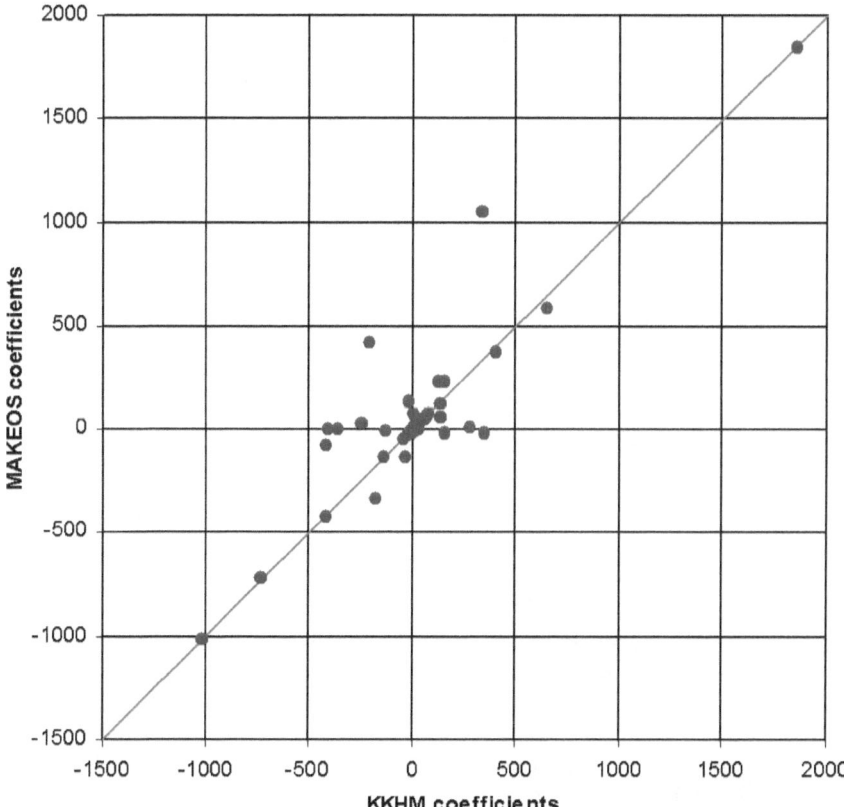

The first 8 coefficients are very close, as these arise directly from the constant volume specific heat at zero density. Typically, the constant pressure specific heat is reported in the literature; however, at zero density, the constant volume specific heat is easily calculated from the following relationship:

$$C_{P0} - C_{V0} = R \qquad (3.21)$$

Data Weighting

Presumably, the other coefficients would also match more closely, if we had used exactly the same data and precision because the analytics are identical. There is also the matter of scaling (or weighting), which we did not discuss in the preceding section on the linear least-squares method. Clearly, some data are more accurate and/or more important than others. This can be accounted for by weighting the data points as they are accumulated.[35] This is already built into the function provided. The inputs to the function are as follows:

```
void LinearLeastSquaresLinearConstraints(double*X,double
    Y,double S,int Na,int Nc,int option)
    {
/* This will fit a user-defined function Y(X) by
    linear-least-squares with linear constraints
    using Lagrange multipliers
    definition of variables
    Na ...... number of adjustable coefficients (Na>1)
    Nc ...... number of constraints (Nc>=0)
    Ne ...... number of constants/equations (Ne=Na+Nc)
    S ....... scaling (or weighting) factor
    X ....... independent functions supplied by user
              for option=2,3 and
              significance levels returned to the
              calling program for option=4
    Y ....... dependent variable supplied by user
              for option=2,3 and the solution for
              option=4 */
```

Haar, Gallagher, and Kell EOS

The next significant advance in the thermodynamic properties of steam came in 1984 out of the National Bureau of Standards (now known as NIST) and the NRC. This is an elegant formulation based on two limiting cases: 1) ideal gas and 2) the Ursell-Mayer virial expansion.[36,37] This relationship arises from quantum theory and approaches the behavior of hard ellipsoids. A perturbation limit is used in this case. Haar, Gallagher, and Kell (HGK)spliced these two extremes together so as to best approximate measured data. One advantage of this approach is more reasonable behavior at elevated pressures, beyond the range of the data used to determine the coefficients. Rather than KKHM's simple powers of density, HGK used functions to approximate Q that vanish on both ends (very low and very high density). These inherently preserve the two limiting cases.[38] The combined expression for HFE can be written:[39]

$$A(\rho,T) = A_{IG}(T) + A_{HD}(\rho,T) + A_{TR}(\rho,T) \qquad (3.23)$$

---

[35] Haar, Gallagher, and Kell provide an excellent discussion of data weighting.

[36] Ursell, H. D., *Proceedings of the Cambridge Philosophical Society*, Vol. 23, p. 685, 1927.

[37] Mayer, J. E., and M. G. Mayer, *Statistical Mechanics*, John Wiley & Sons, Inc., New York, 1940.

[38] I highly recommend this excellent reference (viz., Haar, Gallacher, and Kell).

[39] I have changed the notation and rearranged the ideal gas terms to better fit this text.

where A_{IG} is the ideal gas portion, A_{HD} is the high density portion, and A_{TR} is the transitional portion. The ideal gas portion can be written:

$$\frac{A_{IG}(T)}{RT} = \ln\left(\frac{\rho RT}{P_0}\right) - 1 - \left(\frac{C_1}{T_R} + C_2\right)\ln(T_R) + \sum_{i=3}^{18} C_i T_R^{i-6} \qquad (3.24)$$

The high density portion can be written:

$$\frac{A_{HD}(\rho,T)}{RT} = -\ln(1-y) - \frac{\beta-1}{1-y} + \frac{\alpha+\beta+1}{2(1-y)^2} + 4y\left(\frac{\overline{B}}{b} - \gamma\right) - \frac{\alpha-\beta+3}{2} \qquad (3.25)$$

where $y=b\rho/4$, $\alpha=11$, $\beta=133/3$, $\gamma=7/2$, and $P_0=1.01325$ bar (1 atm.). The two parameters b and B are functions of temperature:

$$b = b_1 \ln\left(\frac{T}{T_0}\right) + \sum_{j=0,1,3,5} b_j\left(\frac{T_0}{T}\right)^j \qquad (3.26)$$

$$\overline{B} = \sum_{j=0,1,2,4} B_j\left(\frac{T_0}{T}\right)^j \qquad (3.27)$$

The transitional portion can be written:

$$\frac{A_{TR}(\rho,T)}{RT} = \sum_{i=1}^{36} \frac{g_i}{k_i}\left(\frac{T_0}{T}\right)^{l_i}\left(1-e^{-\rho}\right)^{k_i} + \sum_{i=37}^{40} g_i \delta_i^{l_i} e^{\left(-\alpha_i\delta_i^{k_i} - \beta_i\tau_o^2\right)} \qquad (3.28)$$

where δ_i and τ_i are defined by:

$$\delta_i = \frac{\rho-\rho_i}{\rho_i}$$
$$\tau_i = \frac{T-T_i}{T_i} \qquad (3.29)$$

HGK provide an excellent discussion of how the data were selected and categorized, especially as it relates to accuracy and importance. An abridged section of their discussion appears below. Notice in particular their discussion of weighting individual data points.

Base Function: Values for the parameters b and B were obtained by nonlinear fits along isotherms to the selected PpT data at 50°K intervals. The equation for B at high values of temperature is in accord with a Stockmayer intermolecular potential. The equation for b at high values of temperature is a linear function of the logarithm of the temperature. This is consistent with theory (assuming a molecular repulsion that increases exponentially with decreasing intermolecular separation). These features ensure that the base function, and hence the complete Helmholtz function, extrapolates in temperature (as well as in density) in accord with molecular theory.

65

Residual Function: The accuracy of the Helmholtz function is in major part controlled by the first 36 terms of the residual function. The coefficients for these were obtained by a least-squares "global" fit to the data, using a linear regression procedure. The fit was made to differences between the selected data and the sum of contributions from the base function and the ideal gas function. Data in the vicinity of the critical point were not included in the global fit. No arbitrary constraints were imposed. That is, the equation was not mathematically constrained to satisfy exactly any thermodynamic condition or any particular datum (e.g., PρT values at the critical point.).

Lastly, the quality of the fit is particularly sensitive to the statistical weight assigned to each datum used for the fit. Values for these weights were obtained from a tedious and painstaking examination of the consistency between data sets, which involved many trial fits. The coefficients for those terms of the residual function that contribute only in the very small, localized regions ($37 < i < 40$), were obtained by least-squares fits to differences between PρT data and the sum of contributions from the base function and from the global terms of the residual function (the terms $1 < i < 36$). The result of these procedures is that the derived Helmholtz function is consistent with trends of data for each data set, but only to within the mutual consistency between different data sets.

We will not repeat either of these tedious processes here (isotherm fitting or data weighting). The source code (nbsnrc.c), along with the input and output files, can be found in the folder examples\NBSNRC. A snipped follows:

```
double bb(double T)
  {
  return(BP[0]+BP[1]*(log(T/To))+BP[3]*pow(To/T,3)
    +BP[5]*pow(To/T,5));
  }
double dbdT(double T)
  {
  return(BP[1]/T-(3.*BP[3]*pow(To/T,4)
    +5.*BP[5]*pow(To/T,6))/To);
  }
double BB(double T)
  {
  return(BQ[0]+BQ[1]*(To/T)+BQ[2]*pow(To/T,2)
    +BQ[4]*pow(To/T,4));
  }
double AAA(double T,double rho)
  {
  int i;
  double Abas,Agas,Ares,b,B,delta,E,tau,y;
  b=bb(T);
  B=BB(T);
  y=b*rho/4.;
  Abas=(-log(1.-y)-(beta-1.)/(1.-y)
        +(alpha+beta+1.)/2./pow(1.-y,2)
        +4.*y*(B/b-gamma)
        -(alpha-beta+3.)/2.
```

66

```
       +log(rho*R*T/Po))*R*T;
  E=exp(-rho);
  for(Ares=i=0;i<36;i++)
    Ares+=GG[i]/kk[i]*pow(To/T,ll[i])*pow(1.-E,kk[i]);
  for(i=36;i<40;i++)
    {
    delta=rho/RR[i-36]-1.;
    tau=T/TT[i-36]-1.;
    Ares+=GG[i]*pow(delta,ll[i])
          *exp(-AA[i-36]*pow(delta,kk[i])-BE[i-
    36]*tau*tau);
    }
  Agas=1.+(CC[0]/(T/100.)+CC[1])*log(T/100.);
  for(i=2;i<18;i++)
    Agas+=CC[i]*pow(T/100.,i-5);
  Agas*=-R*T;
  return(Abas+Ares+Agas+R*(Uref+Sref*T));
  }
double SSS(double T,double rho)
  {
  int i;
  double b,B,delta,E,Sbas,Sgas,Sres,tau,Tdb,TdB,x,y;
  Sgas=1.+CC[0]/(T/100.)+CC[1]*(1.+log(T/100.));
  for(i=2;i<18;i++)
    Sgas+=CC[i]*(i-4)*pow(T/100.,i-5);
  Sgas*=R;
  b=bb(T);
  B=BB(T);
  Tdb=T*dbdT(T);
  TdB=T*dBdT(T);
  y=b*rho/4.;
  x=1.-y;
  Sbas=log(x)+(beta-1.)/x-(alpha+beta+1.)/2./x/x-
    rho*(b+Tdb)*(B/b-gamma)
      +(alpha-beta+1.)/2.-log(rho*R*T/Po)-
    Tdb*rho/4./x+(beta-1.)*Tdb*rho/4./x/x
      -(alpha+beta+1.)*Tdb*rho/4./x/x/x-rho*(TdB-
    B*Tdb/b);
  Sbas*=R;
  E=exp(-rho);
  for(Sres=i=0;i<36;i++)
    Sres+=GG[i]*ll[i]*pow(To/T,ll[i]+1)*pow(1.-
    E,kk[i])/kk[i]/To;
  for(i=36;i<40;i++)
    {
    delta=rho/RR[i-36]-1.;
    tau=T/TT[i-36]-1.;
    Sres-=2.*GG[i]*pow(delta,ll[i])*BE[i-36]*tau
```

```
        *exp(-AA[i-36]*pow(delta,kk[i])-BE[i-
   36]*tau*tau)/TT[i-36];
      }
   return(Sbas+Sres+Sgas-R*Sref);
   }
double PPP(double T,double rho)
   {
   int i;
   double b,B,delta,E,q,Q,tau,y;
   E=exp(-rho);
   for(Q=i=0;i<36;i++)
     Q+=GG[i]*E*pow(1.-E,kk[i]-1)*pow(To/T,ll[i]);
   for(i=36;i<40;i++)
     {
     delta=rho/RR[i-36]-1.;
     tau=T/TT[i-36]-1.;
     Q+=GG[i]*(ll[i]-kk[i]*AA[i-
   36]*pow(delta,kk[i]))*pow(delta,ll[i]-1)
       *exp(-AA[i-36]*pow(delta,kk[i])-BE[i-
   36]*tau*tau)/RR[i-36];
     }
   b=bb(T);
   B=BB(T);
   y=b*rho/4.;
   q=(1.+(alpha+beta*y)*y)/pow(1.-y,3)+4.*y*(B/b-gamma);
   return(rho*(q*R*T+rho*Q));
   }
```

This produces a similar output as KKHM.c...

```
NBS/NRC 1984 Steam Properties
developed by Haar, Gallagher, and Kell
implemented by Dudley J. Benton
  Ts           Ps          Vf          Vg            Hf      Hg
    Sf       Sg
273.160  0.000610716 1.00053 206328.73002      0.0  2497.5
     0.0000 9.1446
275.000  0.000697338 1.00038 181907.95781      7.6  2500.9
     0.0278 9.0958
645.000 21.466270921 2.36766       4.67722  1934.0  2309.2
     4.1710 4.7530
647.126 22.021829771 3.10559       3.10559  2095.8  2095.8
     4.4189 4.4189
   T      P          V          H        S
 275    0.01     1.00038      7.6  0.0278
1600    0.01 73843.28335 5481.1 11.6184
1600 163.84      4.78303 5277.1  6.9911
```

68

Wagner and Pruß EOS

The next significant advancement in the thermodynamic properties of steam came between 1995 and 2002.[30] This effort, sponsored by The International Association for the Properties of Water and Steam (IAPWS), known as the "scientific formulation" and given the designation IAPWS-SF95, is based on temperature and density. The subsequent "industrial formulation," known as IAPWS-IF97, uses temperature and pressure. We will only consider the former.

Wagner and Pruß (WP) took an entirely empirical approach, similar to Keenan, Keyes, Hill, and Moore and distinct from Haar, Gallagher, and Kell. The two formulations (KKHM and WP) differ only in the form and number of terms in the expansion of the zero density and non-ideal components, which minimizes the modifications required to the regression tool (MAKEOS.c) in order to determine the coefficients, should one wish to do so. You can find the source code, along with the input and output files, in the folder examples\SF95. We begin with an expression for the HFE:

$$\frac{a}{RT} = \ln\left(\frac{\rho}{\rho_c}\right) + \varphi(T) + \psi(T,\rho) \tag{3.30}$$

The terms on the right-hand-side of 3.30 are dimensionless. The second term, φ, depends only on temperature (representing the zero-density contribution) and the third term, ψ, depends on temperature and density. The second term is given by the expansion:

$$\varphi = a_1 + a_2\tau + a_3\ln\tau + \sum_{i=4}^{8} a_i \ln\left(1 - e^{-\gamma_i \tau}\right) \tag{3.31}$$

where τ is defined by:

$$\tau = \frac{T_C}{T} \tag{3.32}$$

The third term is given by the expansion:

$$\psi = \sum_{i=1}^{7} g_i \delta^{d_i} \tau^{t_i} + \sum_{i=8}^{51} g_i \delta^{d_i} \tau^{t_i} e^{-\delta^{c_i}} + \\ \sum_{i=52}^{54} g_i \delta^{d_i} \tau^{t_i} e^{-\alpha_i(\delta - \varepsilon_i)^2 - \beta_i(\tau - \gamma_i)^2} + \sum_{i=55}^{56} g_i \Delta^{b_i} \delta\xi \tag{3.33}$$

where δ, Δ, θ, and ζ are defined by:

$$\delta = \frac{\rho}{\rho_C} \tag{3.34}$$

$$\Delta = \theta^2 + B_i\left[(\delta-1)^2\right]^{\alpha_i} \qquad (3.35)$$

$$\theta = (1-\tau) + A_i\left[(\delta-1)^2\right]^{\frac{1}{2\beta_i}} \qquad (3.36)$$

$$\xi = e^{-C_i(\delta-1)^2 - D_i(\tau-1)^2} \qquad (3.37)$$

Snippets from the source code (SF95.c) appear below:

```
static double phi(double delta,double tau)  /* ideal gas
   part */
   {
   int i;
   double t1;
   t1=0.;
   for(i=3;i<8;i++)
     t1+=ai[i]*log(1.-exp(-gi[i]*tau));
   return(log(delta)+ai[0]+ai[1]*tau+ai[2]*log(tau)+t1);
   }
static double psi(double delta,double tau)
   {
   int i;
   double d0,p,p1,p2,t1,t2,t3,t4,theta;
   t1=0.;
   for(i=0;i<7;i++)
     t1+=ar[i]*(pOw(delta,dd[i]))*(pOw(tau,tt[i]));
   t2=0.;
   for(i=7;i<51;i++)
     t2+=ar[i]*(pOw(delta,dd[i]))*(pOw(tau,tt[i]))*exp(-
     pOw(delta,cc[i]));
   t3=0.;
   for(i=51;i<54;i++)
     {
     p1=(delta-1.)*(delta-1.);
     p2=(tau-gg[i-51])*(tau-gg[i-51]);
     t3+=ar[i]*(pOw(delta,dd[i]))*(pOw(tau,tt[i]))*exp(-
     20.*p1-be[i-51]*p2);
     }
   t4=0.;
   for(i=54;i<56;i++)
     {
     theta=(1.-tau)+0.32*pOw((delta-1.)*(delta-
     1.),0.5/be[i-51]);
     d0=theta*theta+0.2*pOw((delta-1.)*(delta-1.),3.5);
     p=exp(-CC[i-54]*((delta-1.)*(delta-1.))-DD[i-
     54]*((tau-1.)*(tau-1.)));
     t4+=ar[i]*pOw(d0,bb[i-54])*delta*p;
     }
   return(t1+t2+t3+t4);
   }
```

```
double P95(double T,double rho)
  {
  double delta,tau;
  delta=rho/rhoc;
  tau=Tcrit/T;
  return(rho*R*T*(1.+delta*psid(delta,tau)));
  }
double S95(double T,double rho)
  {
  double delta,tau;
  delta=rho/rhoc;
  tau=Tcrit/T;
  return(R*(tau*(phit(delta,tau)+psit(delta,tau))-
    phi(delta,tau)-psi(delta,tau)));
  }
double H95(double T,double rho)
  {
  double delta,tau;
  delta=rho/rhoc;
  tau=Tcrit/T;

  return(R*T*(1.+tau*(phit(delta,tau)+psit(delta,tau))+
    delta*psid(delta,tau)));
  }
```

As with **KKHM.c** and **NBSNRC.c**, the main program spits out a table of properties. A brief excerpt follows:

```
IAPWS-SF95 (2002) Steam Properties
developed by Wagner and Pruß
implemented by Dudley J. Benton
  Ts          Ps          Vf          Vg          Hf      Hg
   Sf      Sg
273.160  0.000611657 1.00021 205990.47575     0.0 2500.9
   0.0000 9.1555
647.096 22.064000000 3.10559      3.10559 2084.3 2084.3
   4.4070 4.4070
  T      P        V          H        S
 275    0.01    1.00011      7.8    0.0283
1600  163.84    4.78365   5294.3    7.0156
```

Mollier Chart

Before we leave steam we must consider the Mollier chart, which is a graph of enthalpy vs. entropy. The unique thing about this representation is that the curves are continuous in slope, unlike Z vs. Pr or Tr or Pr vs. Vr. The curves are continuous on the fugacity chart (page 20) but the vapor dome is collapsed to a curve and we don't see the meta-stable states. You will find a spreadsheet containing this (mollier.xls) in the examples\AllSteam folder.

71

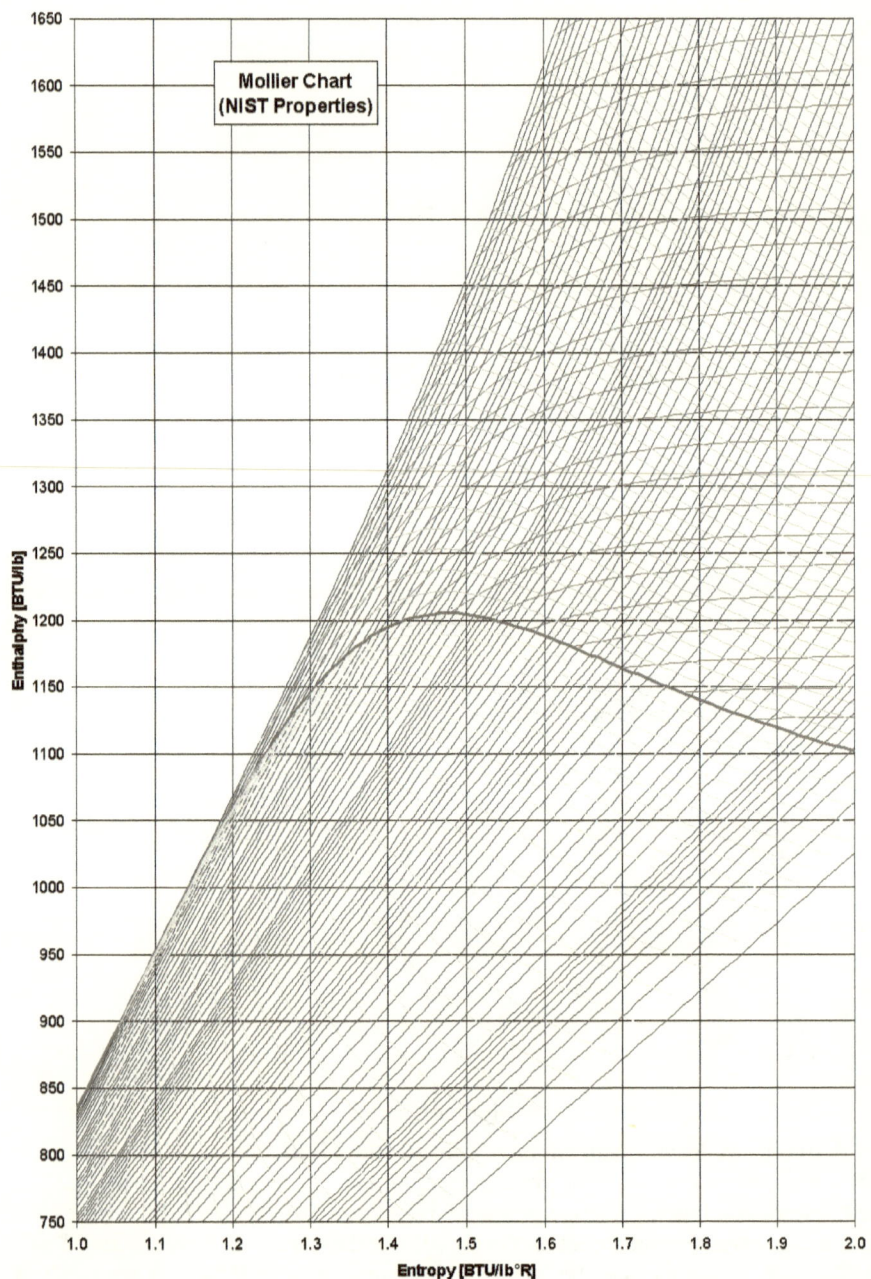

Mollier Chart
(NIST Properties)

Enthalphy [BTU/lb]

Entropy [BTU/lb°R]

Chapter 4. AllSteam Excel® Add-In

We have so far covered the '69, '84, and '95 steam properties, but have only mentioned the '67 and '97 in passing. For completeness, all 5 will be implemented in the Excel® Add-In. You can find the '67 and '97 property functions in the examples\IF67 and examples\IF97 folders, respectively. All of the code for the Add-In can be found in the examples\AllSteam folder. The Add-In is compiled by drawing code from the other folders so that there is only one copy of the functions for each set of properties. Conditional compilation statements (#ifdef _TEST) are used to facilitate the process.

The most extensive text on Excel® Add-Ins is that of Steve Dalton.[40] This excellent document can be found in PDF form at several locations on the Web. Add-Ins (and, for that matter software) development is beyond the scope of this text. The reader is directed to Dalton's text for the former and Kernighan and Ritchie for the latter. The Web is replete with information on software development, including the Wikipedia® K&R site. There is some brief comments in AllSteam.c specific to this end.

An Excel® Add-In is a DLL file renamed to have the extension XLL. It is a pascal wrapper for relocatable object modules. The calling convention and strings must be in pascal form, which is easily accomplished in C. The wrapper registers the functions with calls into Excel®. If this is successful, a spreadsheet can then use the functions inside the wrapper. You must load the Add-In after launching Excel®. The bitness (i.e., either 32-bit or 64-bit) of the Add-In must match the bitness of MS Office®. Note that this may not be the same as the bitness of the O/S (i.e., a 32-bit version of Windows® can only run a 32-bit version of MSO; whereas, a 64-bit version of Windows® can run either a 32-bit or a 64-bit version of MSO). The source code provided can be compiled for either bitness. In fact, the batch file provided produces both.

Academic (Non-Commercial) Use Only

I provide these steam properties and associated Excel® Add-In for academic (i.e., non-commercial) use only. While I have contributed considerable effort to translate these property functions into C and build them into the Add-In wrapper, the underlying code (mostly in FORTRAN) is the work of others (as noted) and can be found on the Web. There are several agents selling similar code for commercial. I do not want to compete with these. I also assume no liability for the use of this software.

---

[40] Dalton, S., Excel Add-in Development in C/C++: Applications in Finance, John Wiley & Sons, Ltd., Chichester, England, 2005.

Excel® Property Functions

The following functions are included:

Hf(Ts,year)
Hg(Ts,year)
HofPS(P,S,year)
HofTP(T,P,year)
Psat(Ts,year)
Sf(Ts,year)
Sg(Ts,year)
SofPH(P,H,year)
SofTP(T,P,year)
TofPH(P,H,year)
TofPS(P,S,year)
Tsat(Ps,year)
Vf(Ts,year)
Vg(Ts,year)
VofTP(T,P,year)

The units are as follows:

property	English	SI
temperature	[°F]	[°K]
pressure	[psia]	[MPa]
specific volume	[ft³/lbm]	[cm³/g]
enthalpy	[BTU/lbm]	[kJ/kg]
entropy	[BTU/lbm/°R]	[kJ/kg/°R]

The year must be one of: 67, 69, 84, 95, or 97. Use +year for English units and −year for SI units.

You will also find a spreadsheet (AllSteam.xls) in this same folder that illustrates the use of the functions. An excerpt appears below:

ASME-1967							
[°K]	[MPa]	[cm³/g]		[kJ/kg]		[kJ/kg/°R]	
Tsat	Psat	Vf	Vg	Hf	Hg	Sf	Sg
273.15	0.0006	0.99536	205301	0.0	2501.6	-0.0002	9.1577
275	0.0007	0.99523	180880	7.8	2505.0	0.0283	9.1086
280	0.001	0.9952	129697	28.8	2514.1	0.1041	8.9800
285	0.0014	0.99556	94258.2	49.7	2523.3	0.1783	8.8572
290	0.0019	0.99626	69380.4	70.7	2532.4	0.2511	8.7397
295	0.0026	0.99725	51686.9	91.6	2541.6	0.3226	8.6274
300	0.0035	0.9985	38946.7	112.5	2550.6	0.3929	8.5199
600	12.349	1.53309	13.6731	1505.9	2681.7	3.5199	5.4795
605	13.172	1.56675	12.5004	1538.8	2662.9	3.5724	5.4305
610	14.037	1.60445	11.3942	1573.1	2641.4	3.6265	5.3780
615	14.947	1.64726	10.3471	1608.9	2616.5	3.6826	5.3212
620	15.906	1.69681	9.35445	1646.8	2587.8	3.7413	5.2593
625	16.914	1.75553	8.40763	1688.0	2554.6	3.8048	5.1915
630	17.977	1.82912	7.48074	1733.8	2514.8	3.8750	5.1145
635	19.096	1.92599	6.56825	1783.0	2466.1	3.9495	5.0250
640	20.277	2.06476	5.62317	1841.2	2401.3	4.0371	4.9121
645	21.523	2.33724	4.46293	1931.4	2292.1	4.1731	4.7323
647.30	22.12	3.15457	3.15457	2107.4	2107.4	4.4429	4.4429

Chapter 5. Refrigerants

While water is technically a refrigerant (ASHRAE designation R718), we will consider these separately. Even though there is considerable economic interest in various refrigerants, the sum total of these considerations pale in comparison to that of steam. It is not surprising that far more effort and expense have been invested in measuring and approximating the thermodynamic properties of steam than refrigerants.

The two best sources of information on refrigerants in general are the ASHRAE[41] and NIST[42]. Refrigerants are most often sold under various trademarked names (e.g., Freon®). The manufacturers of these substances (e.g., Dupont[43]) are more than glad to provide you with information regarding the various properties. We will merely present and manipulate these properties here. The reader is directed to these other resources for background information, including why you should choose one substance over another for a particular application.

Data for 38 common refrigerants, spreadsheets, and code related to our treatment of this subject can be found in the folder examples\refrigerants. A summary appears on the next page. Data for a typical fluid (R13) appears on the following page. Several macros are included in the spreadsheet, including ones to retrieve each of the properties in the summary list from the individual pages. All of the data are in SI units (°K, MPa, g/cm³, kJ/kg, and kJ/kg/°K). The first entry in each table is the triple point.

As is customary for refrigerants, the enthalpy and entropy of the saturated liquid at the triple point have been set to zero. Not all tabulated data follow this convention, but it is done here for consistency. In any event, it's just a constant offset. Also customary for refrigerants, the internal energy is not provided, as these are almost never used in a non-flowing system; thus, enthalpy is the appropriate measure of energy. The critical density (ρ_C) and compressibility (Z_C) will be used in our calculations, as will be the Pitzer acentric factor, w.

---

[41] The American Society of Heating, Refrigerating and Air-Conditioning Engineers is a global professional association seeking to advance heating, ventilation, air conditioning and refrigeration systems design and construction. Much information is available online and in the various editions of the Handbook of Fundamentals. https://www.ashrae.org

[42] The National Institute of Standards and Technology RefProp project is particularly helpful. https://www.nist.gov/srd/refprop

[43] E. I. du Pont de Nemours and Company, commonly referred to as DuPont, is an American conglomerate that was founded in July 1802 in Wilmington, Delaware, as a gunpowder mill by French-American chemist and industrialist Éleuthère Irénée du Pont. https://www.chemours.com/Refrigerants/en_US/products/Freon/

Summary of 38 Refrigerants

Name	Formula	MW 1/mole	Tc °K	Pc MPa	ρc g/cm³	Zc -	R kJ/kg/°K	Tref °K	Pitzer acentric
R11	CCl3F	137.37	471.2	4.409	0.5538	0.2792	0.06053	162.0	0.1915
R12	CF2Cl2	120.93	385.2	4.115	0.5581	0.2785	0.06875	115.4	0.1764
R13	CClF3	104.46	301.5	3.870	0.5779	0.2790	0.07960	122.0	0.1792
R13b1	CBrF3	148.91	340.2	3.964	0.7449	0.2801	0.05584	105.4	0.1722
R14	CF4	88.01	227.7	3.745	0.6257	0.2782	0.09447	89.3	0.1740
R21	CHCl2F	102.92	451.6	5.167	0.5413	0.2616	0.08078	200.0	0.2038
R22	CHClF2	86.48	369.2	4.977	0.5248	0.2672	0.09614	113.2	0.2220
R23	CHF3	70.01	299.1	4.836	0.5251	0.2593	0.11876	118.2	0.2654
R40	CH3Cl	50.49	416.3	6.759	0.3581	0.2754	0.16468	175.4	0.1634
R50	CH4	16.04	190.7	4.641	0.1618	0.2903	0.51836	90.9	0.0112
R113	Cl2FC-CClF2	187.38	487.2	3.392	0.5600	0.2802	0.04437	238.7	0.2518
R114	ClF2C-CClF2	170.92	418.8	3.257	0.5800	0.2756	0.04865	180.4	0.2544
R115	F3C-CClF2	154.47	353.1	3.129	0.6148	0.2678	0.05383	167.0	0.2494
R134a	CH2FCF3	102.03	374.2	4.059	0.5119	0.2600	0.08149	176.5	0.3270
R142b	CH3CClF2	100.50	410.3	4.246	0.4350	0.2876	0.08274	142.0	0.2501
R152a	CHF2CH3	66.05	386.4	4.520	0.3680	0.2525	0.12588	156.2	0.2694
R170	C2H6	30.07	305.4	4.894	0.1930	0.3003	0.27651	92.0	0.0988
R290	C3H8	44.10	369.9	4.251	0.2205	0.2764	0.18855	85.5	0.1523
R500	R12/R152a	99.30	375.3	4.173	0.4970	0.2672	0.08373	128.2	0.2170
R502	R22/R115	111.60	355.3	4.075	0.5606	0.2746	0.07450	167.0	0.2186
R503	R23/R13	87.20	292.7	4.359	0.5640	0.2770	0.09535	122.0	0.1978
R504	R32/R115	79.20	335.3	4.439	0.5500	0.2293	0.10498	129.6	0.2650
R505	R12/R31	103.50	390.9	4.727	0.5368	0.2804	0.08033	150.0	0.1776
R506	R31/R114	93.70	414.8	5.167	0.5510	0.2547	0.08873	158.7	0.2345
R600	C4H10	58.10	425.2	3.797	0.2282	0.2735	0.14311	135.0	0.2015
R600a	C4H10	58.10	408.1	3.648	0.2212	0.2823	0.14311	113.4	0.1848
R702	H2	2.02	33.0	1.293	0.0314	0.3024	4.12424	13.8	-0.2184
R704	He	4.00	5.2	0.248	0.0696	0.3297	2.07758	3.4	-0.3523
R717	NH3	17.03	405.4	11.330	0.2352	0.2434	0.48822	197.2	0.2558
R718	H2O	18.02	647.2	22.089	0.3168	0.2334	0.46150	273.2	0.3438
R720	Ne	20.18	44.4	2.654	0.4831	0.3003	0.41201	24.5	-0.0373
R728	N2	28.01	126.2	3.394	0.3110	0.2913	0.29684	63.2	0.0402
R729	Air	28.97	132.4	3.774	0.3282	0.3026	0.28700	56.9	0.0711
R732	O2	32.00	154.8	5.080	0.4358	0.2899	0.25984	54.4	0.0229
R740	Ar	39.95	150.9	4.898	0.5356	0.2912	0.20813	83.8	-0.0039
R744	CO2	44.01	304.2	7.392	0.4639	0.2773	0.18892	216.5	0.2280
R1150	CH2=CH2	28.05	283.1	5.117	0.2289	0.2664	0.29642	104.0	0.0787
R1270	CH3CH=CH2	42.10	364.9	4.621	0.2203	0.2911	0.19749	88.0	0.1445

Note that the acentric factor for CO_2 is extrapolated.

Thermodynamic Properties of Chlorotrifluoromethane

R13   CClF3   104.459

Ts	log(Ps)	Vf	Zg	Hf	Hg	Sf	Sg
122.0	-3.743	0.5661	0.9993	0.0	180.2	0.0000	1.3628
127.6	-3.394	0.5720	0.9989	4.9	182.6	0.0291	1.3428
133.2	-3.078	0.5781	0.9984	9.8	185.0	0.0582	1.3227
138.7	-2.791	0.5844	0.9975	14.8	187.4	0.0872	1.3026
144.3	-2.525	0.5912	0.9952	19.7	189.7	0.1131	1.2916
149.8	-2.287	0.5980	0.9933	24.6	192.1	0.1453	1.2635
155.4	-2.069	0.6049	0.9906	29.5	194.6	0.1762	1.2386
160.9	-1.868	0.6120	0.9876	34.4	197.0	0.2062	1.2165
166.5	-1.670	0.6193	0.9830	39.4	199.4	0.2332	1.1945
172.0	-1.512	0.6268	0.9788	44.2	201.8	0.2634	1.1798
177.6	-1.352	0.6349	0.9729	49.1	204.0	0.2905	1.1640
183.2	-1.203	0.6437	0.9665	54.2	206.6	0.3171	1.1501
188.7	-1.065	0.6524	0.9588	58.9	208.9	0.3430	1.1377
194.3	-0.936	0.6617	0.9504	63.9	211.2	0.3687	1.1267
199.8	-0.815	0.6711	0.9410	67.7	213.4	0.3976	1.1167
205.4	-0.701	0.6811	0.9305	73.9	215.6	0.4177	1.1079
210.9	-0.594	0.6923	0.9186	79.0	217.8	0.4418	1.0998
216.5	-0.492	0.7036	0.9056	84.1	219.9	0.4655	1.0925
222.0	-0.397	0.7154	0.8913	89.3	221.9	0.4889	1.0859
227.6	-0.306	0.7285	0.8741	94.6	223.8	0.5122	1.0798
233.2	-0.220	0.7423	0.8586	100.0	225.7	0.5352	1.0741
238.7	-0.138	0.7573	0.8402	105.5	226.6	0.5580	1.0687
244.3	-0.060	0.7735	0.8199	111.2	229.1	0.5811	1.0636
249.8	0.015	0.7916	0.7973	117.4	230.5	0.6058	1.0585
255.4	0.086	0.8110	0.7735	123.4	231.9	0.6287	1.0535
260.9	0.154	0.8334	0.7476	129.6	233.0	0.6519	1.0484
266.5	0.219	0.8584	0.7187	135.9	233.9	0.6755	1.0430
272.0	0.282	0.8878	0.6870	142.5	234.3	0.6993	1.0368
277.6	0.343	0.9221	0.6509	149.7	234.5	0.7243	1.0297
280.4	0.373	0.9420	0.6313	153.5	234.3	0.7372	1.0257
283.2	0.402	0.9652	0.6103	157.4	234.0	0.7505	1.0212
285.9	0.430	0.9914	0.5872	161.5	233.5	0.7643	1.0161
288.7	0.459	1.0219	0.5615	165.9	232.6	0.7790	1.0100
291.5	0.486	1.0588	0.5334	170.6	231.3	0.7945	1.0030
294.3	0.514	1.1057	0.5017	175.7	229.6	0.8111	0.9945
297.0	0.541	1.1699	0.4610	181.7	226.8	0.8307	0.9825
299.8	0.567	1.2780	0.3990	189.6	221.2	0.8564	0.9616
301.5	0.588	1.7303	0.2790	205.3	205.3	0.9077	0.9077

Many of these refrigerants are among the fluids used by Nelson and Obert to construct their charts. It is not surprising, therefore, that these exhibit similar behavior for Z vs. Pr, as shown in this next figure:

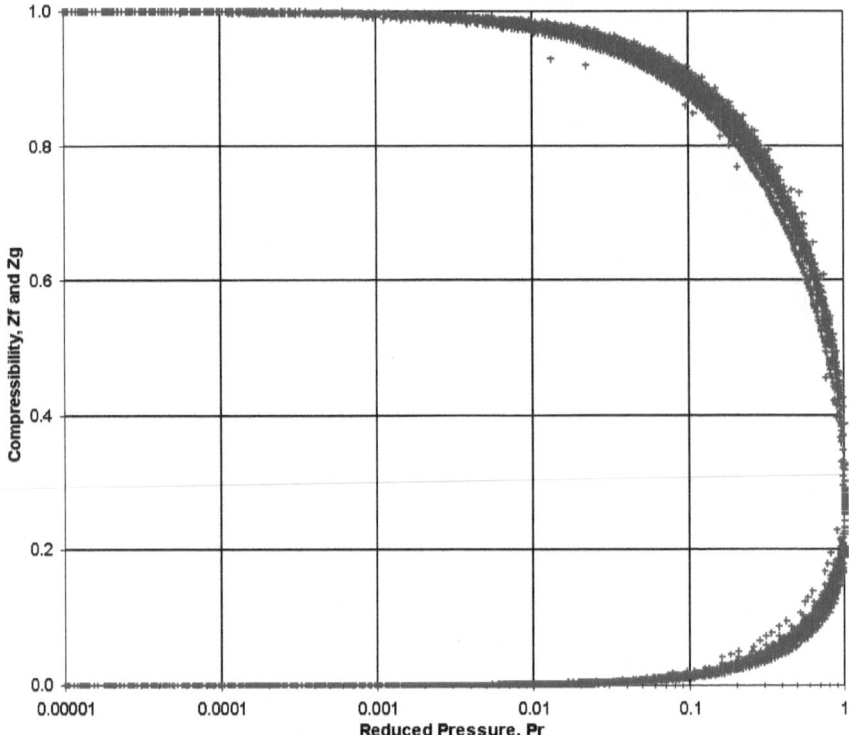

Here we see excellent agreement, except for a few outliers. These "stray" points are from the more unusual fluids, including water and the cryogens. The horizontal axis, log(Pr), tends to squash most of the points to the right side so that the outliers are less noticeable. Of course, some of the data could also be questionable, as these were collected from several sources over many years, some of which have been combined to provide an extended range for a particular fluid. A subsequent graph (log(Ps/Pc) vs. Ts/Tc) will reveal these differences more distinctively.

This next figure is the same data, only plotted vs. reduced temperature:

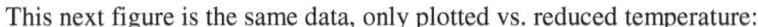

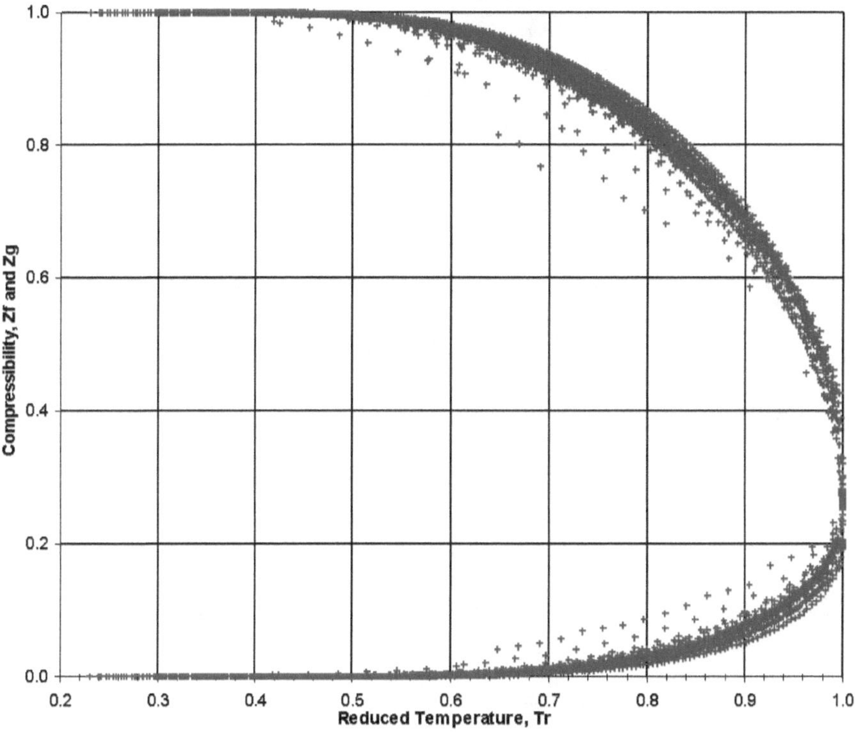

<u>Utility of the Acentric Factor</u>

Pitzer introduced the acentric factor in the context of saturation pressure. That it also is an indication of molecular structure is secondary. Accurately representing saturation pressure is perhaps the first step in approximating the behavior of a substance over the entire range of interest, that is the vapor dome, compressed liquids, and superheated vapors. Most refrigeration cycles operate primarily under the vapor dome, which is why the most common tabulated properties are in this region. There is very little information outside this region for the various refrigerants in the many editions of the ASHRAE Handbook of Fundamentals, which served as the design textbook for decades.

Consider the same summary data, sorted on the acentric factor:

Name	Formula	MW	Zc	Pitzer
		1/mole	-	acentric
R704	He	4.00	0.3297	-0.3523
R702	H2	2.02	0.3024	-0.2184
R720	Ne	20.18	0.3003	-0.0373
R740	Ar	39.95	0.2912	-0.0039
R50	CH4	16.04	0.2903	0.0112
R732	O2	32.00	0.2899	0.0229
R728	N2	28.01	0.2913	0.0402
R729	Air	28.97	0.3026	0.0711
R1150	CH2=CH2	28.05	0.2664	0.0787
R170	C2H6	30.07	0.3003	0.0988
R1270	CH3CH=CH2	42.10	0.2911	0.1445
R290	C3H8	44.10	0.2764	0.1523
R40	CH3Cl	50.49	0.2754	0.1634
R13b1	CBrF3	148.91	0.2801	0.1722
R14	CF4	88.01	0.2782	0.1740
R12	CF2Cl2	120.93	0.2785	0.1764
R505	R12/R31	103.50	0.2804	0.1776
R13	CClF3	104.46	0.2790	0.1792
R600a	C4H10	58.10	0.2823	0.1848
R11	CCl3F	137.37	0.2792	0.1915
R503	R23/R13	87.20	0.2770	0.1978
R600	C4H10	58.10	0.2735	0.2015
R21	CHCl2F	102.92	0.2616	0.2038
R500	R12/R152a	99.30	0.2672	0.2170
R502	R22/R115	111.60	0.2746	0.2186
R22	CHClF2	86.48	0.2672	0.2220
R744	CO2	44.01	0.2773	0.2280
R506	R31/R114	93.70	0.2547	0.2345
R115	F3C-CClF2	154.47	0.2678	0.2494
R142b	CH3CClF2	100.50	0.2876	0.2501
R113	Cl2FC-CClF2	187.38	0.2802	0.2518
R114	ClF2C-CClF2	170.92	0.2756	0.2544
R717	NH3	17.03	0.2434	0.2558
R504	R32/R115	79.20	0.2293	0.2650
R23	CHF3	70.01	0.2593	0.2654
R152a	CHF2CH3	66.05	0.2525	0.2694
R134a	CH2FCF3	102.03	0.2600	0.3270
R718	H2O	18.02	0.2334	0.3438

The molecular weight has nothing to do with the acentric factor, which arises from the relative inter-atomic strain or the configuration of the molecule. The critical compressibility factor, Z_C, and acentric factor, w, do exhibit a weak correlation, as shown in this next figure:

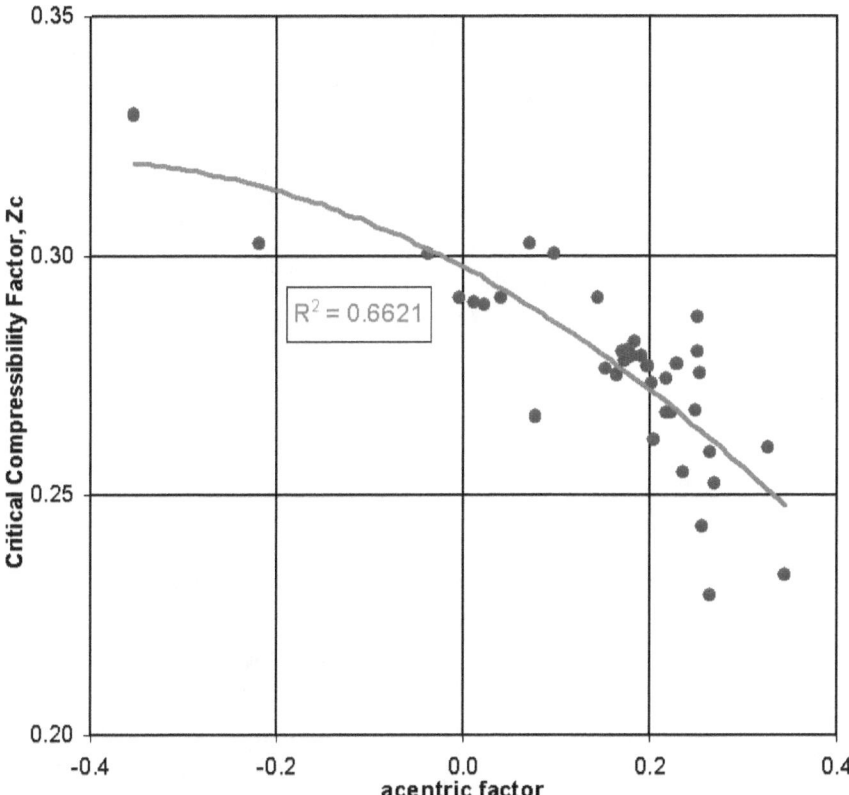

Notice that the cryogenic noble gases (He, H_2, Ne, Ar) are at the top and the polar molecules (CO_2, NH_3, and H_2O) are at the bottom. This sorting confirms the conclusion that the Pitzer acentric factor is a valid measure of molecular eccentricity (or asymmetry).

With this in mind, we will now consider the saturation pressures.

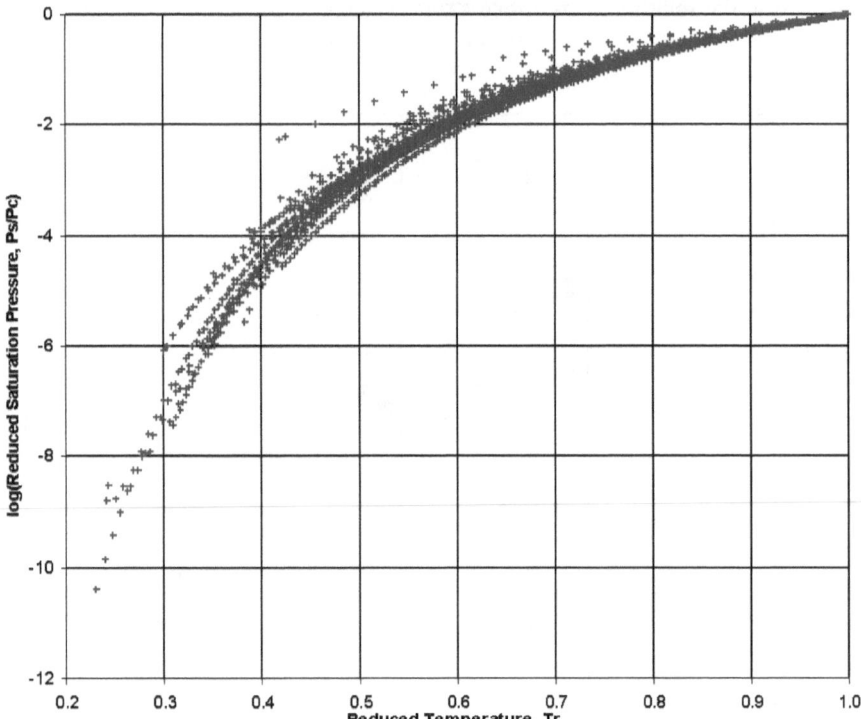

Based on these 38 fluids, we can develop the following empirical relationship, which must be solved implicitly (a macro is provided in the spreadsheet):

$$x = 0.30188\left(\frac{e^x}{T_R^2} - 1\right) - 7.173429\left(\frac{1}{T_R} - 1\right)$$
$$- 2.825214\ln(T_R)$$
$$P_R^{SAT} = e^{x(1+w)}$$

(4.1)

The agreement is acceptable (R^2=0.9966), as illustrated in the following figure:

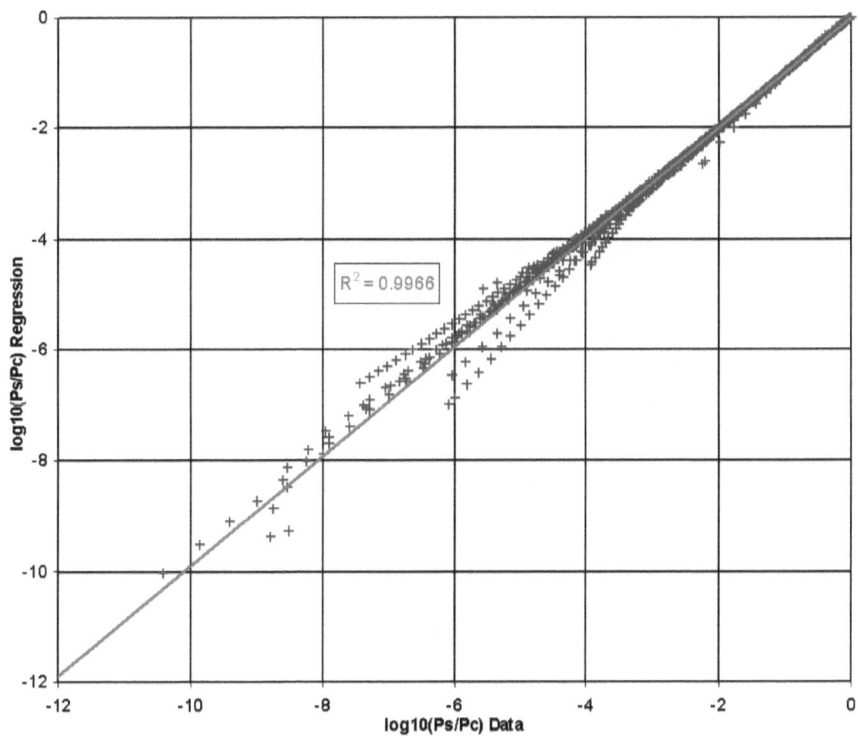

Benedict-Webb-Ruben-Starling EOS

We will use the extended 32-parameter version of the BWR equation of state presented by Starling.[44] This EOS is often the choice of NIST in their development and you can find expanded discussions on their web site as well as many articles on the Web. We will use a modified version of the previous program (MAKEOS.c) to determine the coefficients (BWRSEOS.c). This can also be found in the examples\refrigerants folder.

---

[44] Starling, K. E., *Fluid Properties for Light Petroleum Systems*, Gulf Publishing Company, 1973.

The BWRS expansion for pressure is as follows:

$$P = \sum_{i=1}^{9} a_i(T)\rho^i + \exp(-\delta^2) \sum_{i=10}^{15} a_i(T)\rho^{2i-17}$$

where $\delta = \rho/\rho_c$, and the temperature dependencies of the a_i coefficients are:

$$
\begin{aligned}
a_1 &= RT, \\
a_2 &= b_1 T + b_2 T^{0.5} + b_3 + b_4/T + b_5/T^2, \\
a_3 &= b_6 T + b_7 + b_8/T + b_9/T^2, \\
a_4 &= b_{10} T + b_{11} + b_{12}/T, \\
a_5 &= b_{13}, \\
a_6 &= b_{14}/T + b_{15}/T^2, \\
a_7 &= b_{16}/T, \\
a_8 &= b_{17}/T + b_{18}/T^2, \\
a_9 &= b_{19}/T^2, \\
a_{10} &= b_{20}/T^2 + b_{21}/T^3, \\
a_{11} &= b_{22}/T^2 + b_{23}/T^4, \\
a_{12} &= b_{24}/T^2 + b_{25}/T^3, \\
a_{13} &= b_{26}/T^2 + b_{27}/T^4, \\
a_{14} &= b_{28}/T^2 + b_{29}/T^3, \\
a_{15} &= b_{30}/T^2 + b_{31}/T^3 + b_{32}/T^4.
\end{aligned}
$$

(4.2)

Before we use this EOS on refrigerants, for which we often have only sparse data, we will use it to approximate steam, for which data are plentiful. The first step will be to approximate the critical isotherm ($T=T_C$). This reduces the number of unknowns to 14 (a_2 through a_{15} and γ). Not only do we want an excellent fit to the data, but we also want the smallest possible coefficients to avoid erratic behavior. This can be accomplished using the Excel® Solver (KKHM_BWRS.xls) or with Brent's method (isotherm.c), both in the KKHM folder.

The linear least-squares with linear constraints will assure the best fit that also satisfies 3 constraints at the critical point. We will also consider the magnitude of the coefficients:

$$magnitude = \sqrt{\sum_{i=1}^{14} C_i^2}$$

(4.3)

This next figure shows the magnitude of the coefficients as a function of the exponential multiplier, γ.

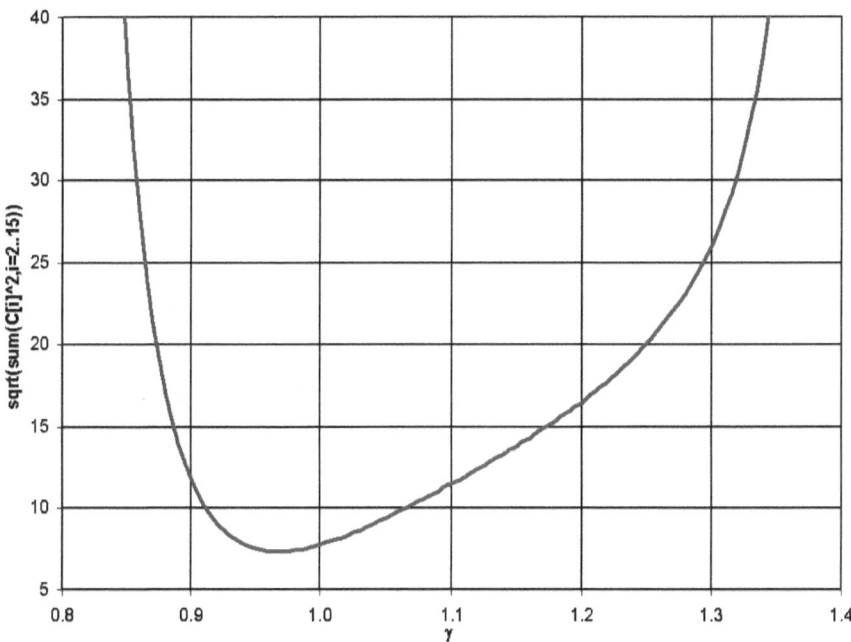

The minimum occurs at $\gamma=0.965$. Agreement with the critical isotherm is excellent. Output of isotherm.c follows…

```
γ=0.85,    √ΣC²=40.4454
γ=0.95,    √ΣC²=7.46214
γ=1.05,    √ΣC²=9.11259
γ=0.995235,  √ΣC²=7.44698
γ=0.973162,  √ΣC²=7.71573
γ=0.97229,   √ΣC²=7.26779
γ=0.961338,  √ΣC²=7.28155
γ=0.967768,  √ΣC²=7.61168
γ=0.96676,   √ΣC²=7.49201
γ=0.962487,  √ΣC²=7.56468
γ=0.964448,  √ΣC²=7.44248
γ=0.965057,  √ΣC²=7.30269
γ=0.965531,  √ΣC²=7.32146
γ=0.964955,  √ΣC²=7.60991
γ=0.965291,  √ΣC²=7.70187
γ=0.965165,  √ΣC²=7.25311
γ=0.965114,  √ΣC²=7.68744
γ=0.96504,   √ΣC²=7.56464
γ=0.965017,  √ΣC²=7.56438
γ=0.965028,  √ΣC²=7.67844
```

```
C[0]=-1.437229770476
C[1]=-3.006526795088
C[2]=3.042680734759
C[3]=-2.604746416272
C[4]=3.943340810760
C[5]=-2.578128294951
C[6]=0.695230477488
C[7]=-0.066597629652
C[8]=2.951097244153
C[9]=0.611513907437
C[10]=-0.085135321249
C[11]=-0.190075906179
C[12]=-0.009712069612
C[13]=-0.008710081116
```

This next figure shows agreement with the critical isotherm.

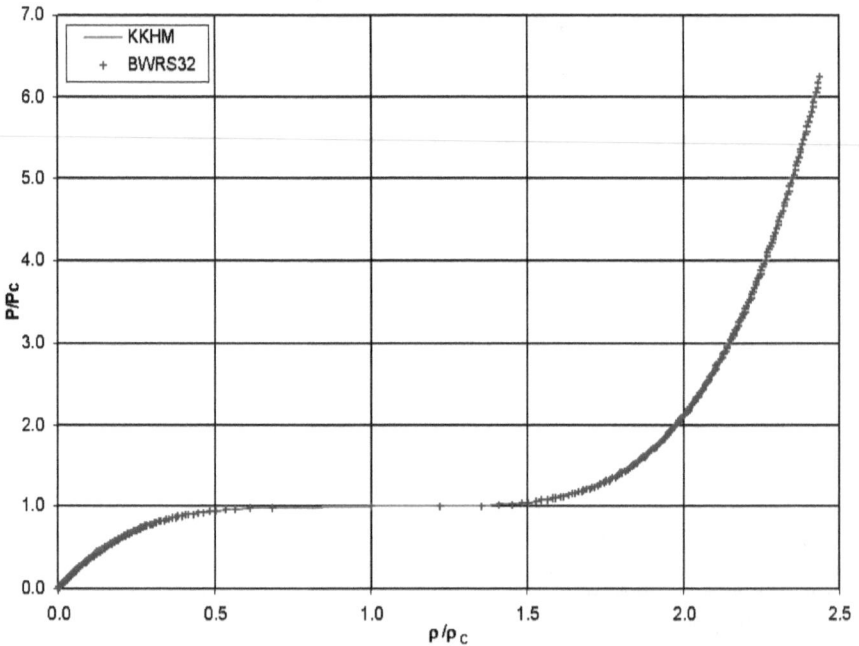

There are a total of 32 coefficients (b_1 through b_{32}) and we have determined 14 of them. These will be implemented as 14 constraints, leaving 18 constants free to minimize the error. This can also be accomplished using the Excel® Solver or C (BWRSpfit.c).

The agreement is acceptable, though not up to the expectations for steam properties. Still, this shows that the EOS is usually adequate for refrigerants.

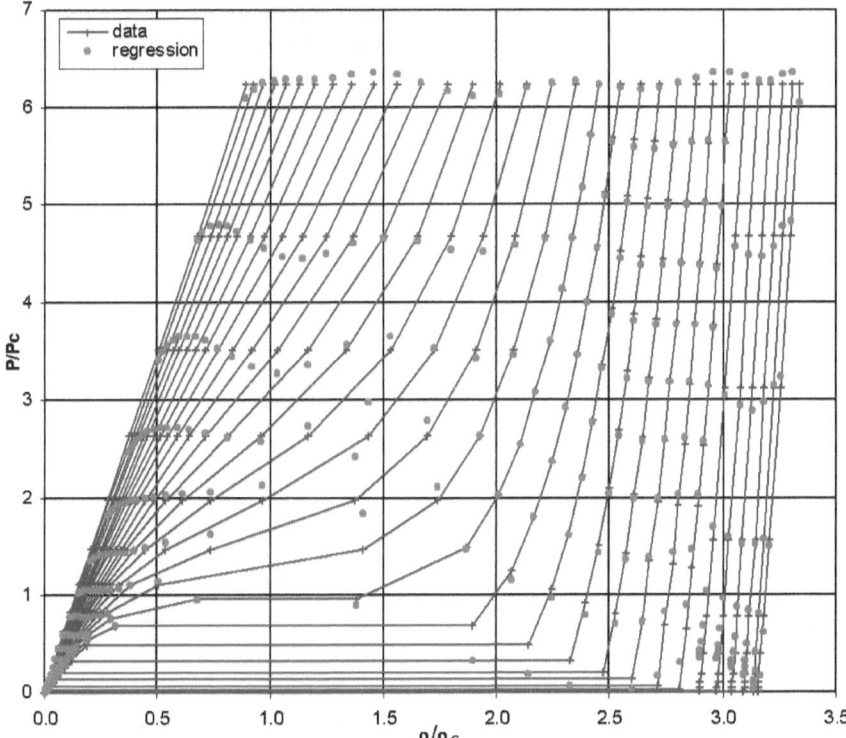

In the case of steam, we controlled the regression by first requiring it to fit the critical isotherm. As it turns out, this step is very important with refrigerants, when the available data may be much more sparse. The program we will use to analyze refrigerants is a modification of MAKEOS.c, called BWRSEOS.c, which you will find in the examples\refrigerants folder.

One fluid that I have recently used this procedure on is NOVEC™ 649, which is a clear, colorless, and low odor fluid, manufactured by 3M™ as a replacement for ozone depleting substances and compounds with high global warming potential.[45] Several recent articles list NOVEC™ 649 and 1230 as having the lowest global warming potential of any available refrigerant. There are two Excel® Add-Ins on my web site that provide properties for this fluid:

https://dudleybenton.altervista.org/miscellaneous/NOVEC649.zip

---

[45] https://multimedia.3m.com/mws/media/569865O/3mtm-novectm-649-engineered-fluid.pdf

The first of these uses the original NIST EOS implemented as VBA® macros and is rather slow, but illustrative. The second uses the 32-parameter BWRS EOS and is implemented in C. In order to check the validity of the regression, BWRSEOS.c creates a polygon output file containing ln(Pr) vs. ln(Vr), such as the following for NOVEC™ 649:

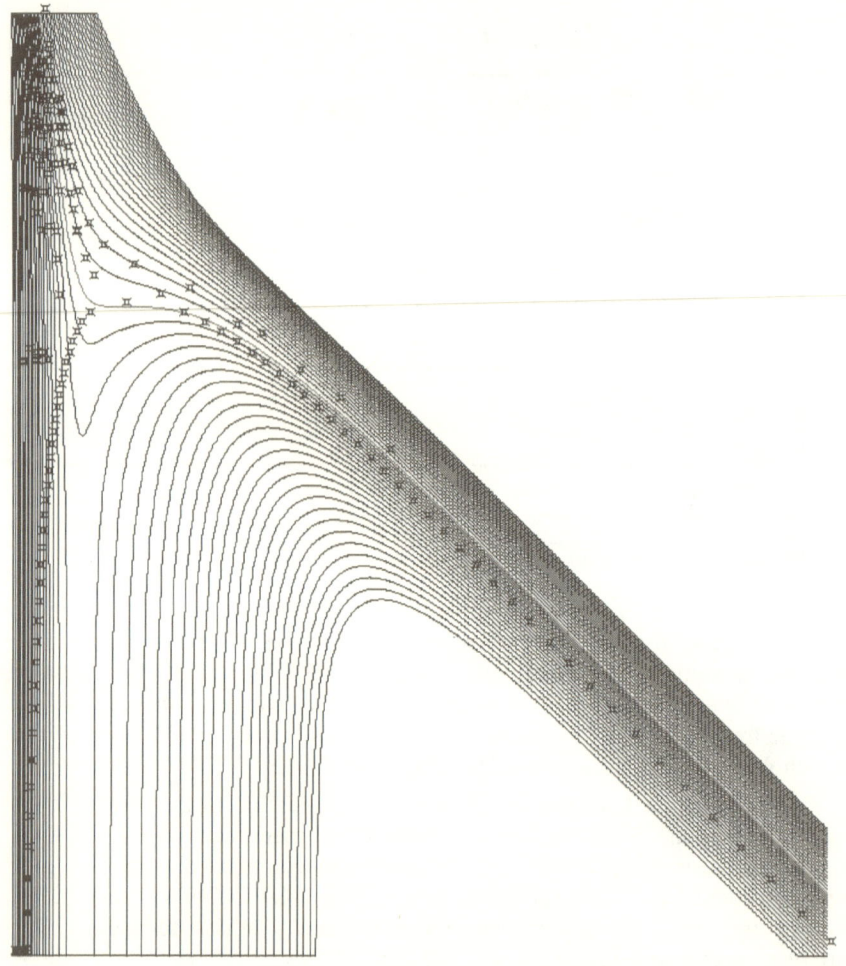

In this case the agreement with data is excellent and the shape is quite satisfactory when compared to the figure on page 3. The asterisks are the data points and the blue curves are the computed isotherms with the red one corresponding to the critical temperature. Notice that the isotherms do not intersect, as illustrated in this next zoomed-in subsection:

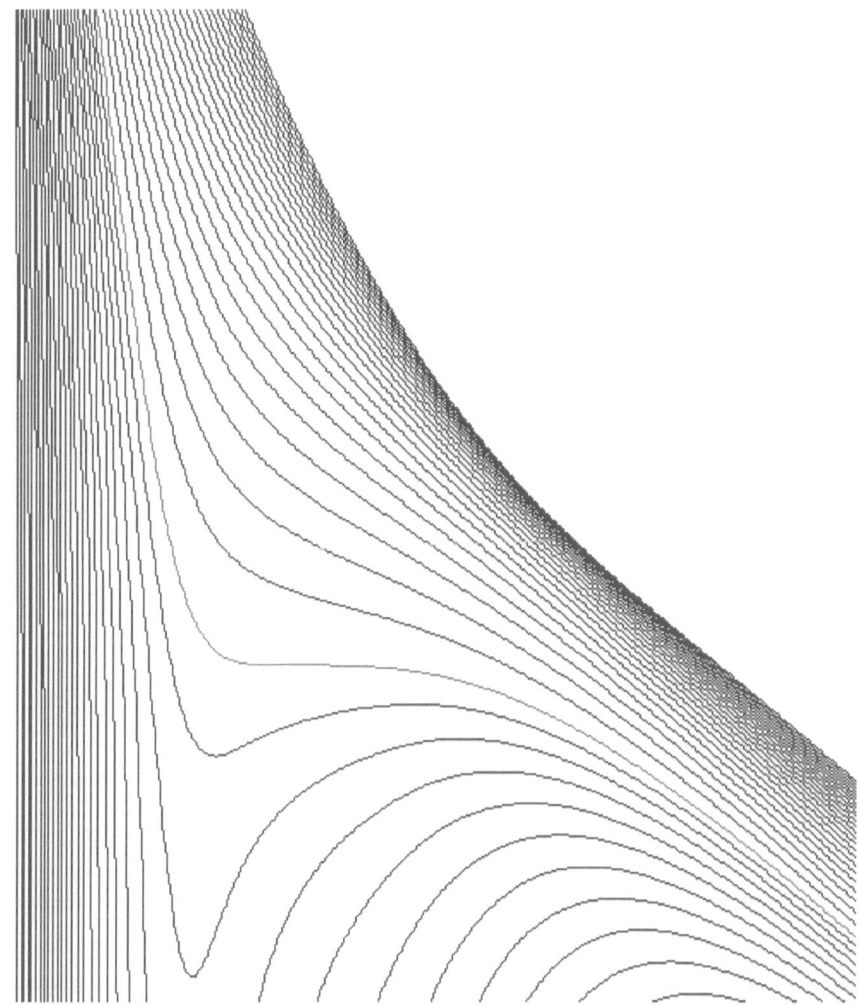

It is essential that the isotherms not intersect, as this would indicate a physically impossible condition, namely two different temperatures at the same density having the same pressure. Temperature and density (or pressure and density) are always independent and uniquely describe the state in equilibrium, unambiguously defining the third state variable (i.e., pressure (or temperature)). The following is one of Maxwell's relations:

$$\left(\frac{\partial P}{\partial T}\right)_v = \left(\frac{\partial s}{\partial v}\right)_T \tag{4.4}$$

Entropy always increases ($\Delta s > 0$) during an isothermal expansion; therefore, the first term in Equation 4.4, $\partial P/\partial T)\rho$=constant, must be positive. The following figure shows the variability for KKHM steam:

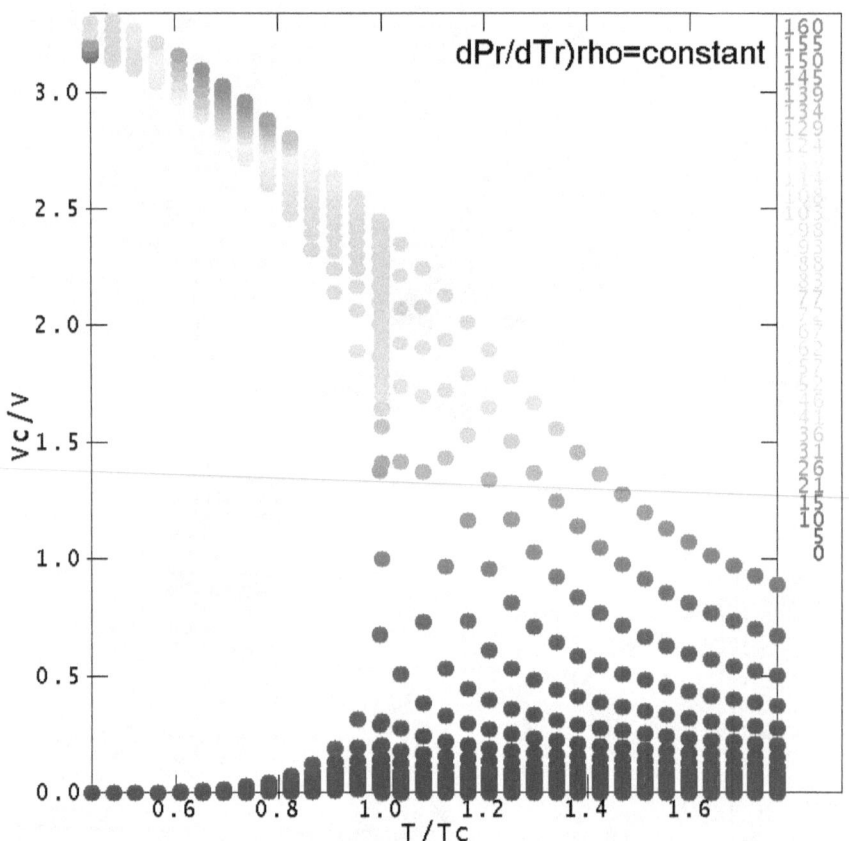

In order to accurately represent thermodynamic properties, we must require this of our constrained regression. There is yet more to be considered before we have a complete picture of the task of representing the thermodynamic properties of fluids... the asymptotic behavior. We will now return to Equation 3.15, which was introduced with the KKHM EOS. In general, the partition function, Q, satisfies the following relationship:

$$Z = 1 + \frac{\partial}{\partial \delta}\left[\delta Q(T_R, \delta)\right] \qquad (4.5)$$

where $T_R = T/T_C$ and $\delta = \rho/\rho_C$. Q also forms the HFE, as in Equation 3.13; therefore, $Q(T_r, 0)$ must be finite, though it need not vanish. We can find Q for BWRS by substituting 4.2 into 4.5 and using Maple® to integrate, which yields:

$$Q(\rho, Tr) = b_1 + \frac{b_2}{\sqrt{Tr}} + \frac{b_3}{Tr} + \frac{b_4}{Tr^2} + \frac{b_5}{Tr^3} + \frac{1}{2}\rho b_6 + \frac{1}{2}\frac{\rho b_7}{Tr} + \frac{1}{2}\frac{\rho b_8}{Tr^2} + \frac{1}{2}\frac{\rho b_9}{Tr^3} + \frac{1}{3}\rho^2 b_{10} + \frac{1}{3}\frac{\rho^2 b_{11}}{Tr}$$

$$+\frac{1}{3}\frac{\rho^2 b_{12}}{Tr^2} + \frac{1}{4}\frac{\rho^3 b_{13}}{Tr} + \frac{1}{5}\frac{\rho^4 b_{14}}{Tr^2} + \frac{1}{5}\frac{\rho^4 b_{15}}{Tr^3} + \frac{1}{6}\frac{\rho^5 b_{16}}{Tr^2} + \frac{1}{7}\frac{\rho^6 b_{17}}{Tr^2} + \frac{1}{7}\frac{\rho^6 b_{18}}{Tr^3} + \frac{1}{8}\frac{\rho^7 b_{19}}{Tr^3}$$

$$-\frac{1}{2}\frac{b_{20}}{\rho Tr^3 \gamma} - \frac{1}{2}\frac{b_{21}}{\rho Tr^4 \gamma} - \left(\left(\frac{1}{2}\frac{1}{\rho Tr^3 \gamma^2} + \frac{1}{2}\frac{\rho}{Tr^3 \gamma}\right)b_{22} - \left(\frac{1}{2}\frac{1}{\rho Tr^5 \gamma^2} + \frac{1}{2}\frac{\rho}{Tr^5 \gamma}\right)b_{23}\right.$$

$$-\left(\frac{1}{\rho Tr^3 \gamma^3} + \frac{\rho}{Tr^3 \gamma^2} + \frac{1}{2}\frac{\rho^3}{Tr^3 \gamma}\right)b_{24} - \left(\frac{1}{\rho Tr^4 \gamma^3} + \frac{\rho}{Tr^4 \gamma^2} + \frac{1}{2}\frac{\rho^3}{Tr^4 \gamma}\right)b_{25}$$

$$-\left(3\frac{1}{\rho Tr^3 \gamma^4} + 3\frac{\rho}{Tr^3 \gamma^3} + \frac{1}{2}\frac{\rho^5}{Tr^3 \gamma}\right)b_{26} + \frac{3}{2}\frac{\rho^3 b_{26}}{Tr^3 \gamma^2} - \left(3\frac{1}{\rho Tr^5 \gamma^4} + 3\frac{\rho}{Tr^5 \gamma^3} + \frac{1}{2}\frac{\rho^5}{Tr^5 \gamma}\right)b_{27}$$

$$(4.6)$$

$$+\frac{3}{2}\frac{\rho^3 b_{27}}{Tr^5 \gamma^2} - \left(12\frac{1}{\rho Tr^3 \gamma^5} + 12\frac{\rho}{Tr^3 \gamma^4} + 2\frac{\rho^5}{Tr^3 \gamma^2} + 6\frac{\rho^3}{Tr^3 \gamma^3} + \frac{1}{2}\frac{\rho^7}{Tr^3 \gamma}\right)b_{28}$$

$$-\left(12\frac{1}{\rho Tr^4 \gamma^5} + 12\frac{\rho}{Tr^4 \gamma^4} + 2\frac{\rho^5}{Tr^4 \gamma^2} + 6\frac{\rho^3}{Tr^4 \gamma^3} + \frac{1}{2}\frac{\rho^7}{Tr^4 \gamma}\right)b_{29}$$

$$-\left(60\frac{1}{\rho Tr^3 \gamma^6} + 60\frac{\rho}{Tr^3 \gamma^5} + 30\frac{\rho^3}{Tr^3 \gamma^4} + \frac{1}{2}\frac{\rho^9}{Tr^3 \gamma} + \frac{5}{2}\frac{\rho^7}{Tr^3 \gamma^2} + 10\frac{\rho^5}{Tr^3 \gamma^3}\right)b_{30}$$

$$-\left(60\frac{1}{\rho Tr^4 \gamma^6} + 60\frac{\rho}{Tr^4 \gamma^5} + 30\frac{\rho^3}{Tr^4 \gamma^4} + \frac{1}{2}\frac{\rho^9}{Tr^4 \gamma} + \frac{5}{2}\frac{\rho^7}{Tr^4 \gamma^2} + 10\frac{\rho^5}{Tr^4 \gamma^3}\right)b_{31}$$

$$-\left.\left(60\frac{1}{\rho Tr^5 \gamma^6} + 60\frac{\rho}{Tr^5 \gamma^5} + 30\frac{\rho^3}{Tr^5 \gamma^4} + 10\frac{\rho^5}{Tr^5 \gamma^3} + \frac{1}{2}\frac{\rho^9}{Tr^5 \gamma} + \frac{5}{2}\frac{\rho^7}{Tr^5 \gamma^2}\right)b_{32}\right) \bigg/ e^{(\gamma \rho^2)}$$

The $\ln(\rho)$ terms cancel out because $A_1 = RT$, but there are still $1/\rho$ terms in 4.6; therefore, these must be set to zero. This forms three more constraints:

$$\gamma^5 b_{20} + \gamma^4 b_{22} + 2\gamma^3 b_{24} + 6\gamma^2 b_{26} + 24\gamma b_{28} + 120 b_{30} = 0$$
$$\gamma^5 b_{21} + 2\gamma^3 b_{25} + 24\gamma b_{29} + 120 b_{31} = 0 \qquad (4.7)$$
$$\gamma^4 b_{23} + 6\gamma^2 b_{27} + 120 b_{32} = 0$$

effectively reducing the number of unknowns to 11 for the BWRS EOS. These are all linear constraints, so implementing them with the code already introduced is straightforward, though not trivial.[46]

---

[46] At this point I must comment about the Maple® software. I can't praise it highly enough. I am glad I spent years solving such equations as this the hard way, but I'm also glad that I never need do it again. I highly recommend this tool, but also knowing how the calculus is done before relying on it to do the heavy lifting.

Importance of the Critical Isotherm

This next figure compares 3 critical isotherms: Nelson_Obert, NOVEC™ 649, and **KKHM** steam.

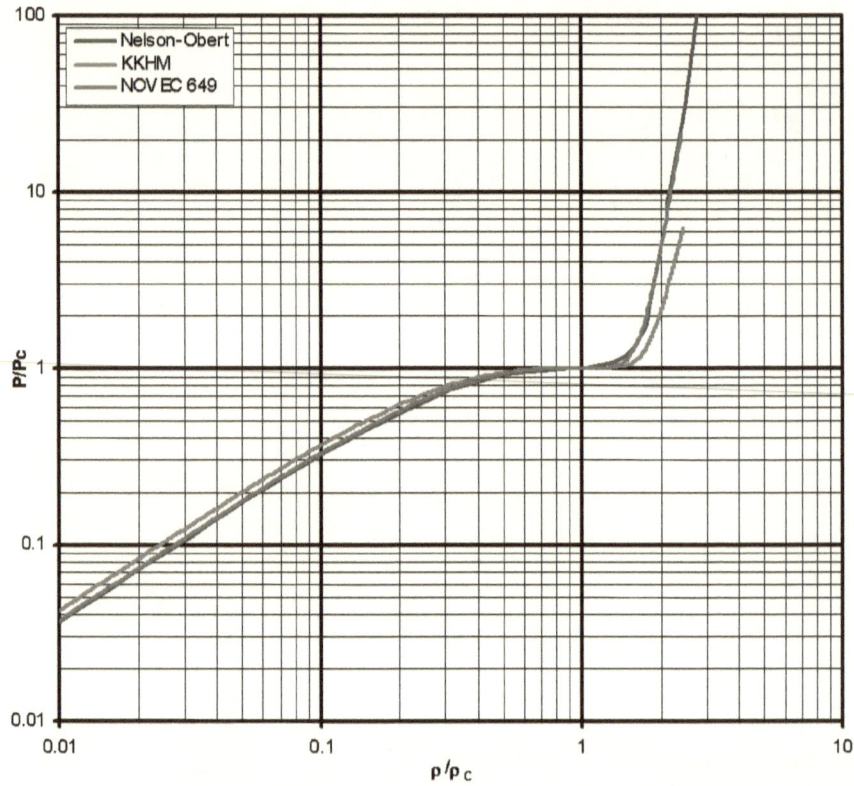

The N-O and NOVEC™ curves are barely distinguishable; whereas, the one for steam is noticeably different. This is another example of how fluids having significantly different critical compressibilities and Pitzer acentric factors can't be treated the same.

NASA Glenn Gas Properties

As in the previous chapter with steam, the complete set of thermodynamic properties requires data for the zero-density constant-volume specific heat. For this, we will turn to the NASA Glenn Report.[47] I have transferred all of the data for thousands of gases into a combined spreadsheet (NASA_Glenn_Gases.xls),

---

[47] McBride, B. J., Zehe, M. J., Gordon, S., "NASA Glenn Coefficients for Calculating Thermodynamic Properties of Individual Species," NASA Report No. 211556, 2002.

which you will find in the folder examples\NASA Glenn. You can find an Excel® Add-In on my web page that facilitates use of this data:

https://dudleybenton.altervista.org/software/NASAGlenn100.zip

Chapter 6. Liquid Metals

References for the thermodynamic properties of mercury abound on the Web. There are numerous hits on a Google™ search. Sadly, these are neither consistent nor complete. The nearest source of complete properties is attributed to Lucian A. Sheldon of the General Electric Corporation circa 1948, which is quite dated.[48] The saturation pressure is well-defined by Huber, Laesecke, and Friend of NIST.[49] Table 2 of this reference lists the span of critical properties over the year 1912-1996, including temperatures ranging from 1543K to 1923K and pressures ranging from 154 to 343.2 MPa. Lucian did not provide critical values, only temperatures up to 1028K.

Mercury makes an excellent case-in-point to gather and build a complete set of thermodynamic properties. The first value required is the molecular weight of the most stable or predominant isotope (or mixture thereof). This is generally agreed to be 200.59. Huber, Laesecke, and Friend (HLF) provide $T_C=1764K$ and $P_C=167$ MPa, along with the vapor pressure curve. This also yields a Pitzer acentric factor of -0.190. The final critical property ($Z_C=0.39$) is provided by Kulinskii.[50] The critical density is then calculated to be $\rho_C=6.1503$ g/cm³. The data and Psat macro can be found in examples\metals\Mercury.xls.

We next construct the Z_F/Z_G vs. P_S/T_S curve, which will be based on this data and also informed by other such curves, including the Nelson-Obert. Low temperature data is provided by Mozaffaria,[51] which was gathered from several references. The slope of the ρ_f vs. Tsat curve as it approaches T_C comes from N-O and also steam in that $1-\rho_c/\rho_f$ is linear with $(1-T_R)^{1/3}$. You will find a macro in the spreadsheet to calculate this curve. The Z_G curve comes from the refrigerant database, that is, a curve-fit of Z_G vs. T_R as a function of Z_C. There is also a macro for this. The end result is a smooth curve that fits data on both ends as well as the middle.

[48] Sheldon, L. A., "Thermodynamic Properties of Mercury Vapor," Transactions of the ASME, Vol. 49A, No. 30, 1949.

[49] Huber, M. L., Laesecke, A., and Friend, D. G., "The Vapor Pressure of Mercury," NIST Report IR-6643, 2006.

[50] Kulinskii, V. L., "The Critical Compressibility Factor Value: Associative Fluids and Liquid Alkali Metals," *Journal of Chemical Physics*, Vol. 141, 2014.

[51] Mozaffaria, F., "Equation of State for Mercury," Journal of Physical Chemistry and Electrochemistry Vol. 1, pp. 37-44, 2010. Note that in spite of the title of this paper, it does not provide an EOS or even a complete set of saturated data, but it is quite worth reading and can be found online.

The constructed saturation compressibility curve for mercury:

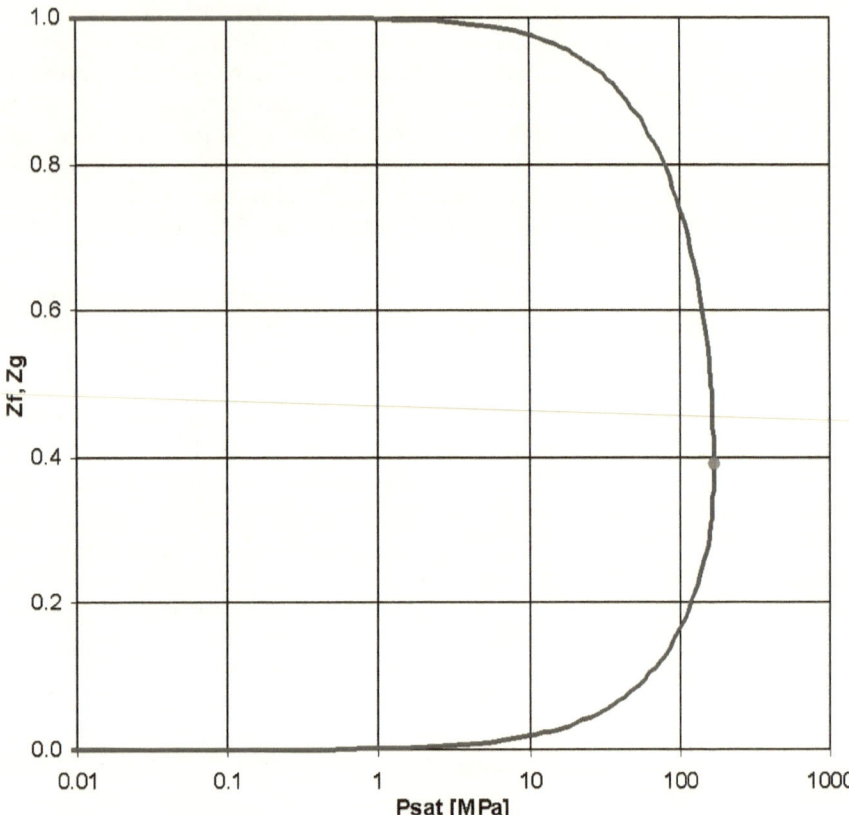

Equation 2.6 is adequate for our purposes. The critical constants are defined by Z_C and the two partial derivatives at the critical point.

The residual volume, B, must lie within the blue area of the following figure, as not all values have a real, positive solution.

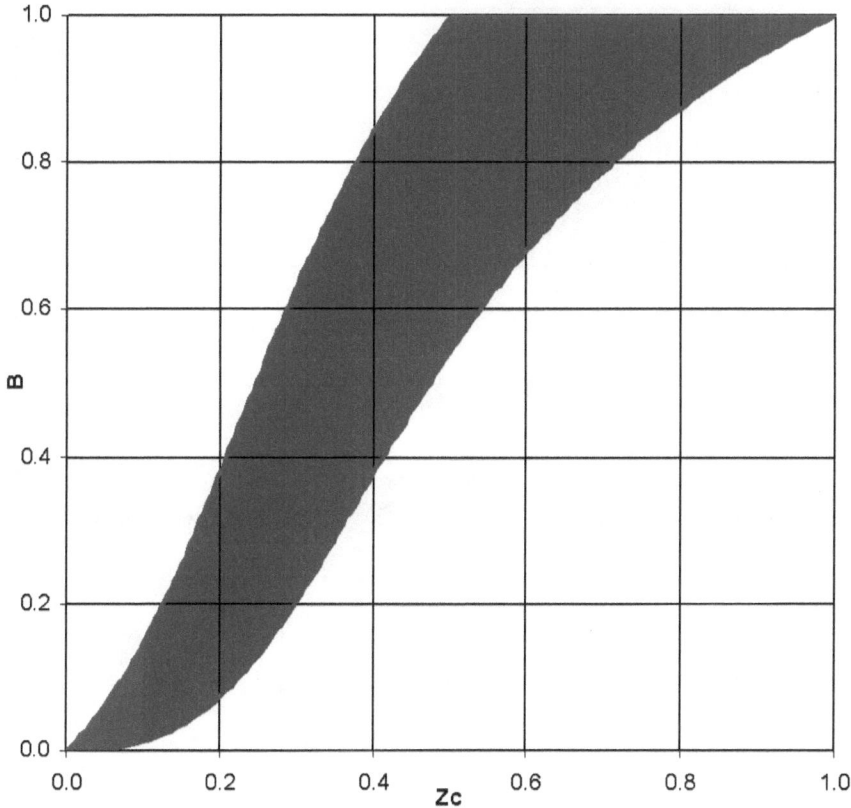

The temperature variation of A, C, and D are found to best fit the Z_F/Z_G curve using the Excel® Solver. The following PV curves are the result:

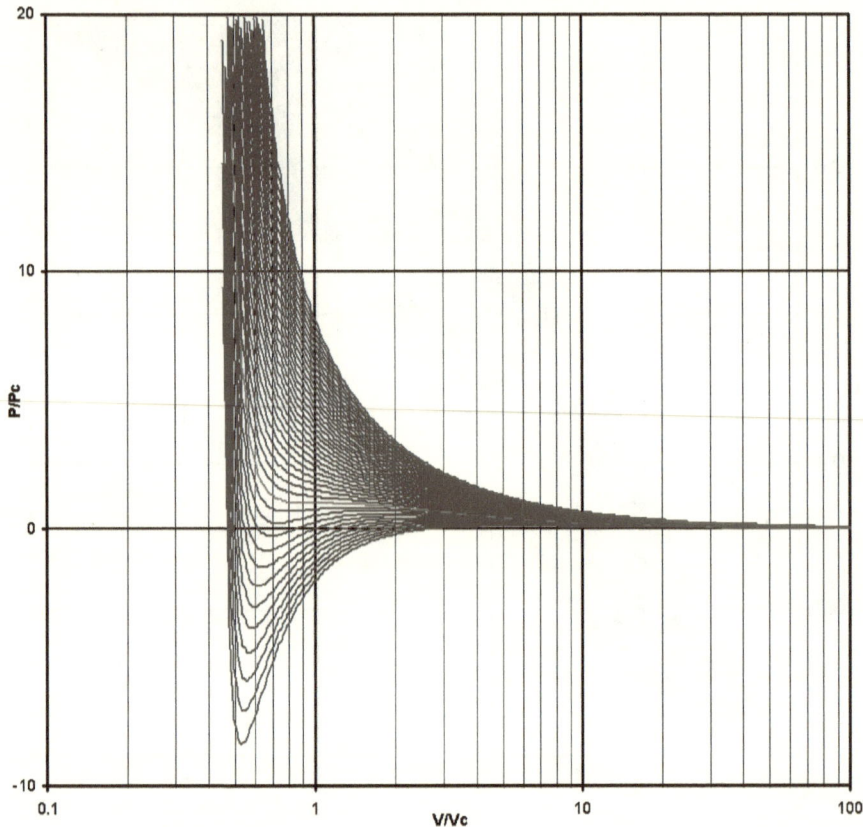

<u>Clapeyron Equation and Latent Heat of Vaporization</u>

The following is one of Maxwell's relations:

$$\left(\frac{\partial s}{\partial v}\right)_T = \left(\frac{\partial p}{\partial T}\right)_v = -\frac{\partial^2 a}{\partial T \partial v} \tag{6.1}$$

The change from saturated liquid to saturated vapor at constant pressure is also at constant temperature; therefore Equation 6.1 can be written:

$$\frac{dp_{sat}}{dT_{sat}} = \frac{s_{fg}}{v_{fg}} \tag{6.2}$$

We also have the following relationship for temperature:

$$T = \left(\frac{\partial h}{\partial s} \right)_p = \left(\frac{\partial u}{\partial s} \right)_v \qquad (6.3)$$

This can be applied across the vapor dome at constant T_{SAT}:

$$h_{fg} = T_s s_{fg} \qquad (6.4)$$

Equation 6.4 is substituted into 6.1 to obtain the Clapeyron equation:

$$\frac{dp_{sat}}{dT_{sat}} = \frac{h_{fg}}{T_{sat} v_{fg}} \qquad (6.5)$$

What this means practically in our development of the properties of mercury is that, if we have the right P_{SAT} curve and the right shape of the Z_F/Z_G curve, then we have the correct latent heat of vaporization, h_{FG}. This can be verified by example, but is unnecessary, as the consequence flows inherently from the calculus.

We get the specific heat from the NASA Glenn spreadsheet already mentioned. It is a trivial matter to integrate CpdT and (Cp/T)dT, then curve-fit the result. This supplies the temperature portion of the non-residual enthalpy and entropy, respectively. The residual portions are supplied by Equations 1.20 and 1.21, respectively. Maple® is used to analytically integrate the EOS.

We will follow this same process for sodium and potassium. The key properties of sodium come from Sobolev[52] as well as Fink and Leibowitz[53] and Dunning.[54] Key properties of potassium come from Heimel[55] as well as Coe[56] and Ewing et al.[57] These documents are readily available online.

With minor modifications, the AllSteam.c code can be adapted to create an Excel® Add-In for these three liquid metals. The functions are the same (see the table on page 74), except that the year is replaced by "Hg", "Na", or "K" to indicate the fluid. Only SI units are provided this time. Should you need unit

---

[52] Sobolev, V., "Database of Thermophysical Properties of Liquid Metal Coolants for GEN-IV: Sodium, Lead, and Bismuth," Scientific Report of the Belgian Nuclear Research Centre, CEN-BLG-1069, 2010 (revised 2011).

[53] Fink, J. K. and Leibowitz, L., "Calculation Of Thermophysical Properties Of Sodium," Argonne National Laboratory Report CONF-8106164-5, 1983.

[54] Dunning, E. L., "The Thermodynamic and Transport Properties Of Sodium and Sodium Vapor,"
Argonne National Laboratory Report ANL-6426, 1960.

[55] Heimel, S., "Thermodynamic Properties of Potassium to 2100°K," NASA Technical Note D-4165, 1967.

[56] Coe, H. H., "Summary of Thermophysical Properties of Potassium," NASA Technical Note D-3120, 1965.

[57] Ewing, C. T., Stone, J. P., Spann, J. R., Steinkuller, E. W., Williams, D. D., and Miller, R. R., "High Temperature Properties of Potassium," Naval Research Laboratory Report 6233, 1965.

conversions, these can readily be done inside Excel®. The code (Metals.c) along with batch files to compile and update the Add-Ins can be found in the examples\metals folder. There is also a spreadsheet that illustrates the use of the functions (Metal_Addin.xls). The saturation table appears below:

Hg   <set this to Hg, Na, or K

Ts	Ps	Vf	Vg	Hf	Hg	Sf	Sg
K	MPa	cm³/g	cm³/g	kJ/kg	kJ/kg	kJ/kg/K	kJ/kg/K
234.3	2.98E-10	0.07615	3.26E+10	0.0	318.7	0.000	1.360
310.8	7.16E-07	0.07680	1.80E+07	7.9	326.7	0.041	1.067
387.3	7.61E-05	0.07751	2.11E+05	15.5	334.6	0.072	0.896
463.8	0.001692	0.07826	1.14E+04	23.1	342.5	0.097	0.786
540.3	0.01536	0.07906	1.46E+03	30.6	350.4	0.119	0.711
616.7	0.07972	0.07991	3.20E+02	38.1	358.2	0.137	0.656
693.2	0.2854	0.08083	1.00E+02	45.6	365.8	0.154	0.615
769.7	0.7890	0.08183	40.026	53.1	373.1	0.168	0.583
846.2	1.811	0.08291	18.995	60.9	379.9	0.182	0.558
922.7	3.623	0.08409	10.218	68.8	386.1	0.194	0.537
999.2	6.53	0.08540	6.0316	77.2	391.3	0.206	0.519
1075.6	10.86	0.08687	3.8177	86.0	395.5	0.217	0.503
1152.1	16.94	0.08852	2.5480	95.4	398.5	0.228	0.489
1228.6	25.11	0.09043	1.7708	105.6	400.0	0.238	0.476
1305.1	35.71	0.09268	1.2691	116.9	399.7	0.249	0.464
1381.6	49.06	0.09540	0.93042	129.5	397.3	0.259	0.452
1458.1	65.48	0.09880	0.69279	144.0	392.4	0.270	0.440
1534.5	85.24	0.10329	0.51997	161.2	384.0	0.282	0.427
1611.0	108.6	0.10973	0.38936	182.4	370.6	0.296	0.412
1687.5	135.8	0.12059	0.28479	211.2	348.7	0.313	0.394
1764.0	167.0	0.17075	0.17075	286.8	286.8	0.355	0.355

table of specific volumes

specific volume cm³/g			
T/P	0.1	1	10	100
300	0.1	0.1	0.1	0.1
400	0.1	0.1	0.1	0.1
500	0.1	0.1	0.1	0.1
3400	1409.3	140.9	14.1	1.4
3500	1450.7	145.0	14.5	1.4
3600	1492.2	149.2	14.9	1.5
3700	1533.6	153.3	15.3	1.5
3800	1575.1	157.5	15.7	1.6
3900	1616.5	161.6	16.1	1.6
4000	1658.0	165.8	16.6	1.6

table of enthalpies

enthalpy [kJ/kg]				
T/P	0.1	1	10	100
300	6.8	6.9	8.3	21.9
400	16.8	17.0	18.3	31.9
500	26.7	26.8	28.2	41.9
600	36.4	36.6	37.9	51.7
700	366.8	46.3	47.7	61.5
800	377.2	376.1	57.5	71.3
900	387.6	386.6	67.5	81.3
1000	397.9	397.1	77.8	91.5
3200	626.0	625.8	624.1	607.4
3300	636.4	636.2	634.6	618.9
3400	646.7	646.6	645.0	630.3
3500	657.1	657.0	655.5	641.7
3600	667.5	667.3	665.9	653.0
3700	677.8	677.7	676.4	664.2
3800	688.2	688.1	686.8	675.4
3900	698.6	698.5	697.3	686.5
4000	709.0	708.8	707.7	697.6

table of entropies

entropy [kJ/kg/K]				
T/P	0.1	1	10	100
300	0.0359	0.0361	0.0383	0.0605
400	0.0767	0.0769	0.0786	0.0952
500	0.1080	0.1081	0.1094	0.1227
600	0.1334	0.1335	0.1346	0.1455
700	0.6599	0.1549	0.1559	0.1651
800	0.6738	0.5773	0.1744	0.1824
900	0.6860	0.5898	0.1909	0.1979
1000	0.6969	0.6009	0.2060	0.2121
3200	0.8176	0.7221	0.6262	0.5266
3300	0.8208	0.7253	0.6295	0.5302
3400	0.8239	0.7284	0.6326	0.5336
3500	0.8269	0.7314	0.6356	0.5369
3600	0.8298	0.7343	0.6386	0.5401
3700	0.8326	0.7372	0.6414	0.5432
3800	0.8354	0.7399	0.6442	0.5462
3900	0.8381	0.7426	0.6469	0.5491
4000	0.8407	0.7453	0.6496	0.5519

Recommendations for Future Development

Before we leave equations of state and take on viscosity, consider again the figure on page 46 showing isotherms of ZTr vs. Vc/V. That figure was generated using the van der Waals EOS. All cubic EOSs produce similar graphs. The corresponding figure for **KKHM** steam appears below:

KKHM Steam

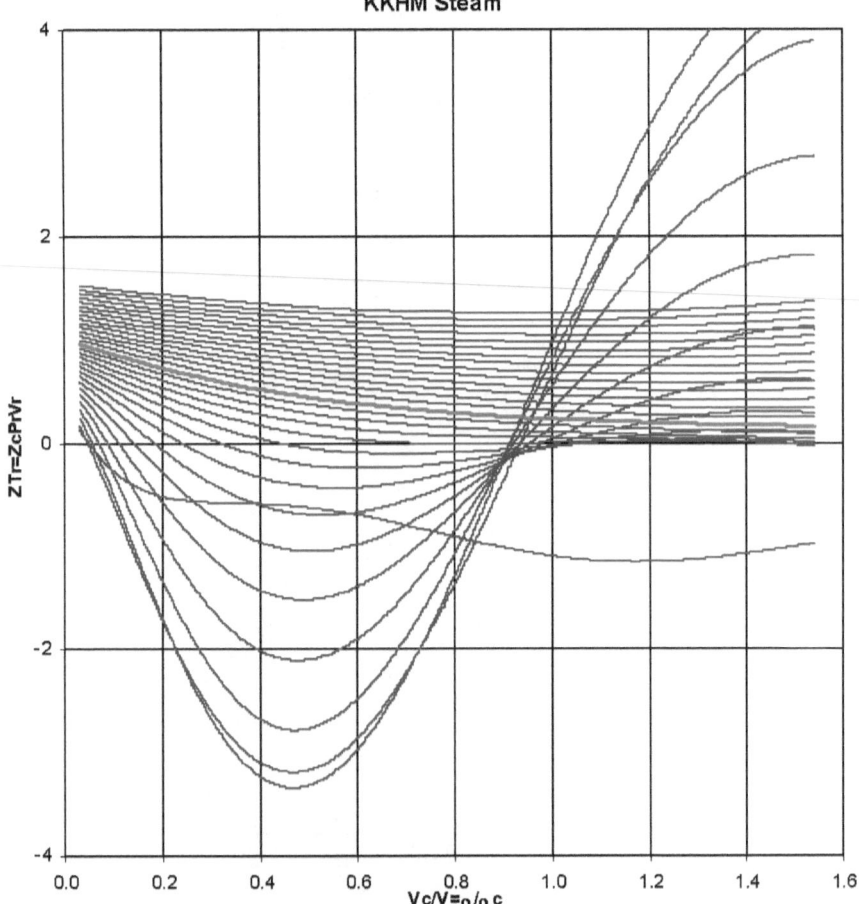

This can be found in a spreadsheet in the examples\KKHM folder. A similar graph can be found in the NBSNRC folder. The corresponding one in the SF95 folder is a real mess. It should be clear from the discussion contained herein that the sub-critical isotherms must have the following form:

$$P = Ps + A(\rho - \rho_F)(\rho - \rho_G)(\rho - \rho_H)\prod(\rho + a_i^2)$$ (6.6)

where $\rho_F > \rho_H > \rho_G$ are the three and only three real positive roots. Maxwell's criterion (Equation 1.17) controls the area under the curve. These wild excursions can't occur if there are only three real positive roots (i.e. under the vapor dome) and the area under the curve is fixed. Any other roots must be either negative or imaginary. The isotherms in the preceding figure must not cross. These are not burdensome or unreasonable requirements for developing future equations of state.

Chapter 7. Viscosity

As with equations of state, we begin with the more general properties, then turn to steam. The most extensive information available apart from water is for natural gas. A variable mixture adds an interesting dimension.

Lee-Gonzalez-Eakin Equation

Perhaps the simplest non-trivial example was published in the Journal of Petroleum Technology and is readily available online.[58] The expression for dynamic viscosity follows the previous work of Starling and Ellington.[59]

$$\mu = Ke^{X\rho^Y} \tag{7.1}$$

$$K = \frac{(7.77 + 0.0063M)T^{1.5}}{122.4 + 12.9M + T} \tag{7.2}$$

$$X = 2.57 + \frac{1914.5}{T} + 0.0095M \tag{7.3}$$

$$Y = 1.11 + 0.04X \tag{7.4}$$

where ρ is the density in gm/cc, T is the temperature in °R, M is the molecular weight, and μ is the dynamic viscosity in micropoise. This formula is implemented as a macro in the spreadsheet natural_gas_viscosity.xls in folder examples\ viscosity. Estimating the properties of a natural gas mixture requires several details for the constituents. These have already been copied into the spreadsheet and may also be found in critical_data.xls. As this formula requires the density, we will use the Peng-Robinson EOS—a common choice for approximating natural gas. The Lee, Gonzalez, and Eakin (LGE) experimental data are also included in the spreadsheet for a check.

---

[58] Lee, A. L., Gonzalez, M. H., and Eakin, B. E., "The Viscosity of Natural Gases," Journal of Petroleum Technology, Vol. 18, No. 8, pp. 997–1000, 1966.
[59] Starling, K. E. and Ellington, R. T., "Viscosity Correlations for Nonpolar Dense Fluids," AIChE Journal, Vol. 11, No. 15, 1964.

LGE considered four mixtures. The spreadsheet has four tabs, one gas mixture per tab. The pseudo critical properties are calculated on each tab and the results used in the property calculations, as illustrated in the following table:

Lee, Gonzalez, and Eakin Gas 1			MW	Tc	Pc	Vc	R	Zc	Pitzer	s.g.
constituent	formula	mole frac		°K	MPa	cm³/g	J/K	-	acentric	air
Methane	CH4	86.30%	16.04	190.9	4.599	6.143	0.5183	0.2855	0.0108	0.554
Ethane	C2H6	6.80%	30.07	305.3	4.872	4.838	0.2765	0.2792	0.0972	1.038
n-Propane	C3H8	2.40%	44.10	369.8	4.244	4.539	0.1886	0.2763	0.1515	1.522
i-Butane	C4H10	0.43%	58.12	407.8	3.640	4.457	0.1431	0.2781	0.1852	2.007
n-Butane	C4H10	0.48%	58.12	425.1	3.798	4.389	0.1431	0.2741	0.1981	2.007
i-Pentane	C5H12	0.11%	72.15	460.4	3.381	4.239	0.1152	0.2701	0.2286	2.491
n-Pentane	C5H12	0.11%	72.15	469.7	3.370	4.308	0.1152	0.2682	0.2510	2.491
n-Hexane	C6H14	0.10%	86.18	507.5	3.012	4.270	0.0965	0.2627	0.2990	2.975
n-Heptane	C7H16	0.04%	100.20	540.3	2.736	4.270	0.0830	0.2606	0.3483	3.460
n-Octane	C8H18	0.00%	114.23	568.8	2.487	4.308	0.0728	0.2587	0.3978	3.944
n-Nonane	C9H20	0.00%	128.26	594.7	2.280	4.326	0.0648	0.2559	0.4425	4.428
n-Decane	C10H22	0.00%	142.28	617.7	2.100	4.389	0.0584	0.2553	0.4881	4.913
Carbon Dioxide	CO2	3.20%	44.01	304.1	7.377	2.135	0.1889	0.2741	0.2667	1.519
Nitrogen	N2	0.00%	28.01	126.2	3.396	3.203	0.2968	0.2904	0.0370	0.967
Oxygen	O2	0.00%	32.00	154.6	5.043	2.291	0.2598	0.2877	0.0216	1.105
Helium	He	0.00%	4.00	5.2	0.227	14.296	2.0773	0.3013	0.0000	0.000
Hydrogen	H2	0.00%	2.02	33.2	1.297	32.200	4.1244	0.3048	-0.2216	0.000
Carbon Monoxide	CO	0.00%	28.01	132.9	3.494	3.627	0.2968	0.3213	0.0477	1.000
Hydrogen Sulfide	H2S	0.00%	34.08	373.4	8.963	2.878	0.2440	0.2832	0.0948	1.000
Water Vapor	H2O	0.00%	18.02	647.1	22.064	3.109	0.4615	0.2297	0.3445	1.000
Argon	Ar	0.00%	39.95	150.8	4.870	1.873	0.2081	0.2906	0.0010	1.000
mixture	mix	99.97%	19.18	209.7	4.685	5.865	0.4785	0.2844	0.0308	0.662

The viscosity calculations are illustrated in the next table:

P	T	ρ	AGA8	P-R	exp.	rep.	calc.	P-R
psia	°F	gm/cc	gm/cc	gm/cc	µp	µp	µp	µp
14.7	160					112		
200	160	0.0108	0.0094	0.0086		115		128
300	160							
400	100	0.0220	0.0217	0.0199		118		118
600	100	0.0344	0.0335	0.0308		122		121
700	100	0.0401	0.0397	0.0366	126.1		124	123
800	100	0.0461	0.0460	0.0425	127.8	127	126	125
1000	100	0.0614	0.0592	0.0548	133.1	133	132	129
2000	100	0.1348	0.1311	0.1205	171.8	176	170	161
2500	100	0.1686	0.1652	0.1515	199.2	203	194	181
3000	100	0.2029	0.1946	0.1787		236		203
3500	100							
4000	100	0.2473	0.2389	0.2225		291		245
5000	100	0.2991	0.2699	0.2553		338		285
6000	100					374		
8000	100	0.3304	0.3263	0.3194		439		389

Experimental and calculated viscosity are compared in this next figure:

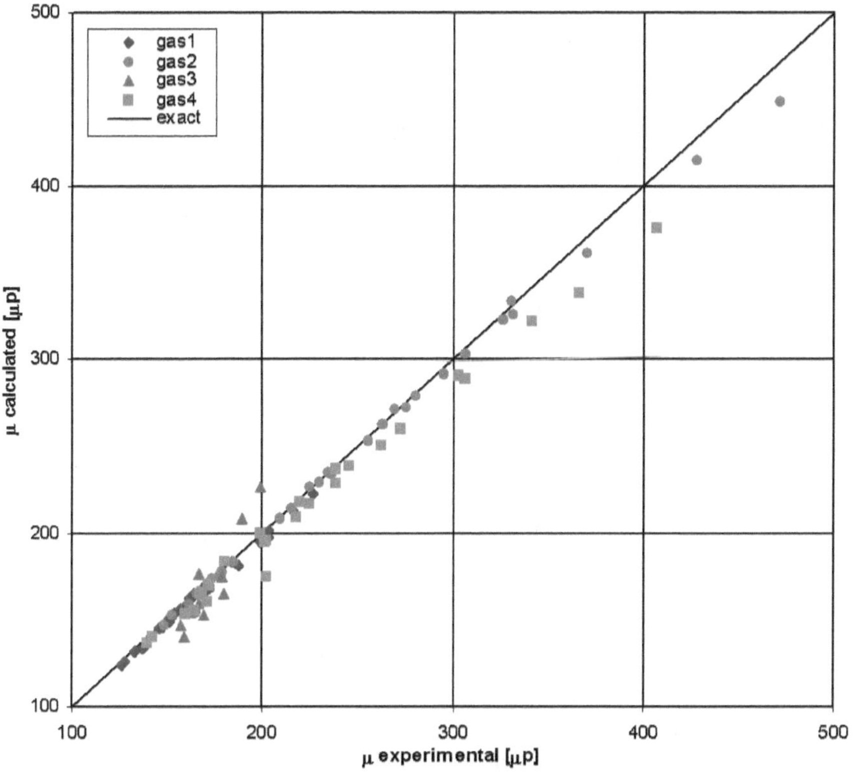

The above results are calculated using the reported (presumably experimental) densities. The calculations using Peng-Robinson densities are also contained in the spreadsheet. The AGA8 densities are also included. AGA8 is the last word on natural gas densities.[60] This formulation is well documented and available online; therefore, it will not be duplicated here. You can even get a spreadsheet to implement the calculations from NIST at the following link:

https://pages.nist.gov/AGA8/

[60] Starling, K. E. and Savidge, J. L., "Compressibility Factors of Natural Gas and Other Related Hydrocarbon Gases," American Gas Association Transmission Measurement Committee Report No. 8, 1992 (revised 1994).

Reported (presumably measured) and Peng-Robinson densities are compared in this next figure:

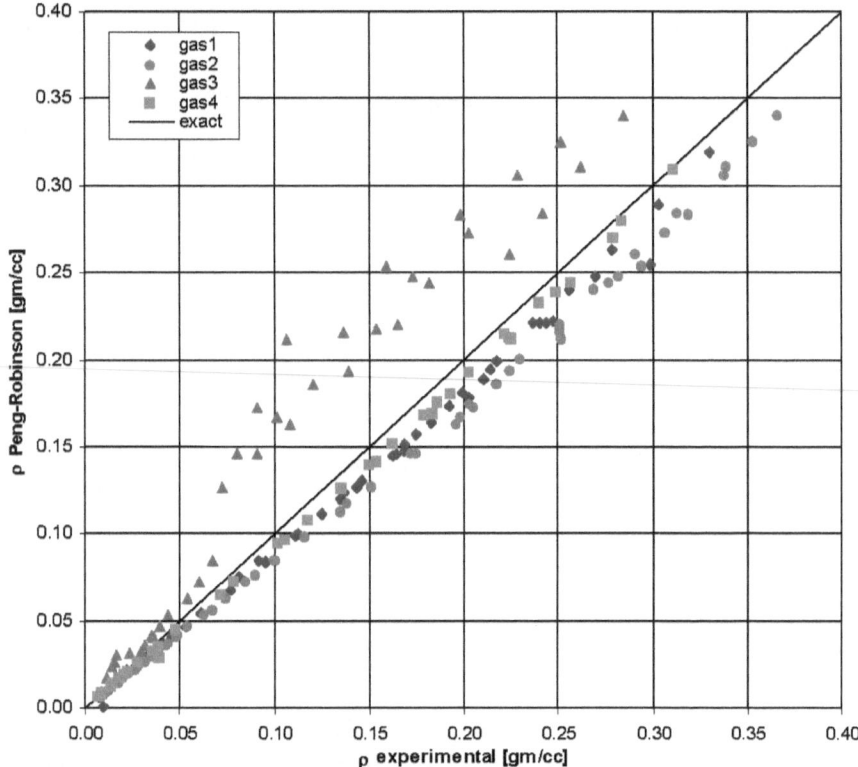

The agreement is reasonable for gases 1, 2, and 4. The disagreement with gas 3 increases with increasing density (and pressure).

The following figure shows the same comparison only with AGA8:

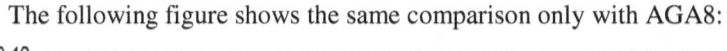

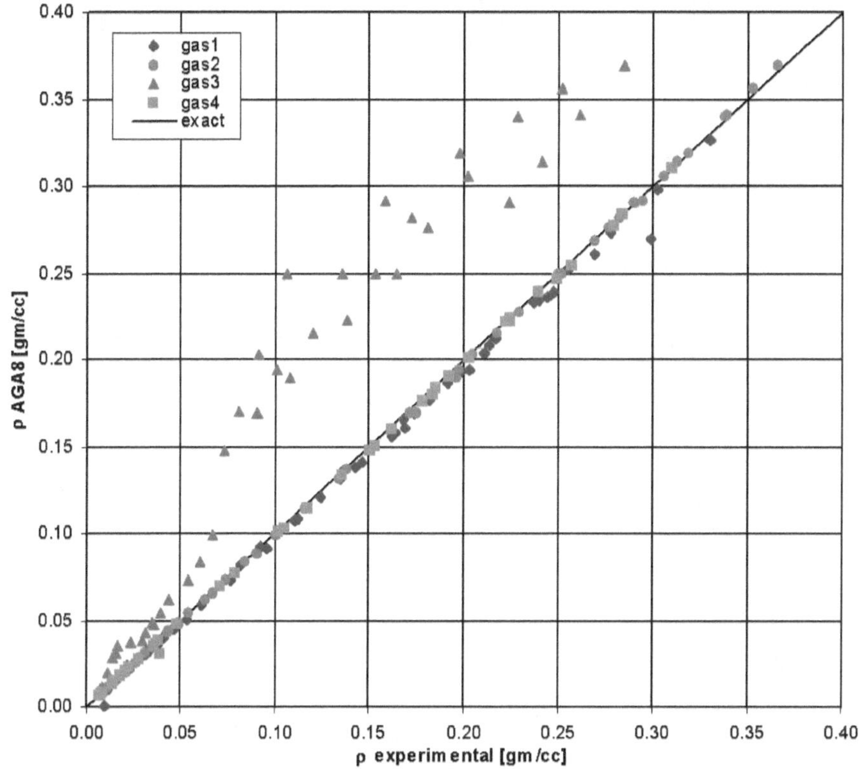

Again gases 1, 2, and 4 are right on the mark, but gas 2 is significantly off. It is extremely unlikely that the AGA8 results are in error; therefore, the reported densities for gas 3 must be incorrect. The fact that P-R is consistently and slightly low (as shown in the preceding figure) is why so much effort has been invested in AGA8.

Miller's Residual Viscosity

Another empirical formula for natural gas viscosity can be found in the Flow Measurement Engineering Handbook.[61] The approach is similar to residual enthalpy and entropy in that it starts with the ideal gas behavior plus a term to account for the departure from ideality. The following code implements this calculation:

```
Function Miller(Tabs,Pcr,Tcr,vCr,Greal,density)
```

---

[61] Miller, R. W., *Flow Measurement Engineering Handbook 2nd Ed.*, McGraw-Hill, 1989.

```
r=density*vCr
a=(1.023+0.23364*r+0.58533*r^2-
  0.40758*r^3+0.093324*r^4)^4
b=(Tcr^(1/6))/Sqr(Greal*28.9625)/
  ((Pcr/0.101325)^(2/3))
Tr=Tabs/Tcr
If(Tr>1)Then
  ucp=3.5*(Tr^(0.71+0.29/Tr))/b
Else
  ucp=3.5*(Tr^0.965)/b
EndIf
Miller=ucp+(a-1)/b
End Function
```

All of the requisite properties, including the specific gravity are calculated
in the spreadsheet. Agreement with experimental data is comparable, as shown
in this next figure:

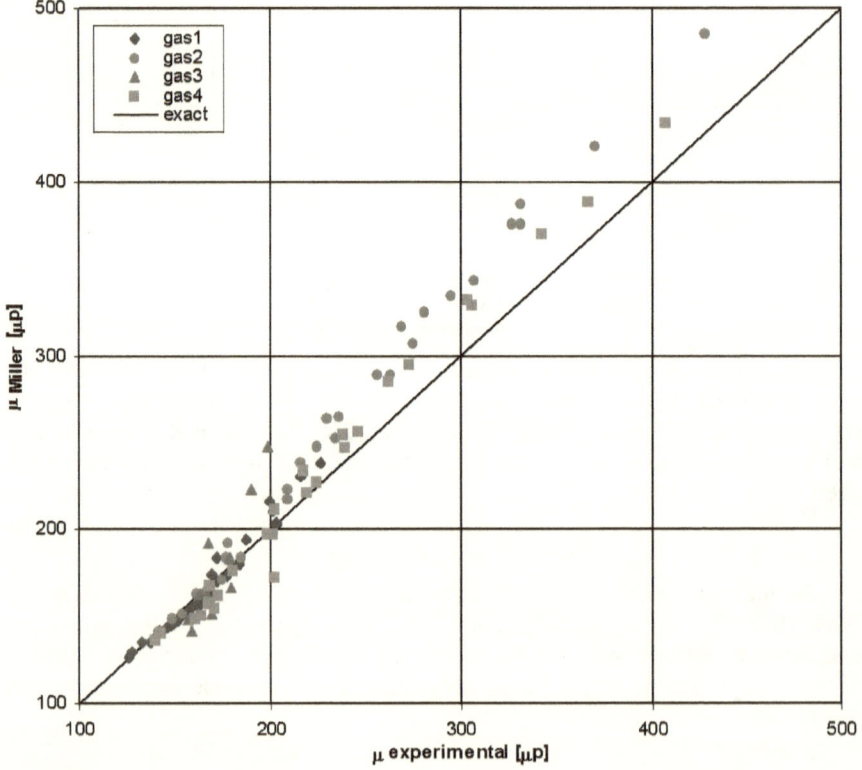

API Viscosity Correlation

A far more complicated empirical relationship for natural gas viscosity can be found in the American Petroleum Institute (API) Handbook.[62] This lengthy correlation involves contribution factors for each of the constituents. I have included these results in column K of the spreadsheet (open without updating to see values). In spite of the increased complexity, the results are no better than the LGE or Miller correlations, as shown in this next figure, which is why I haven't bothered to provide the code.

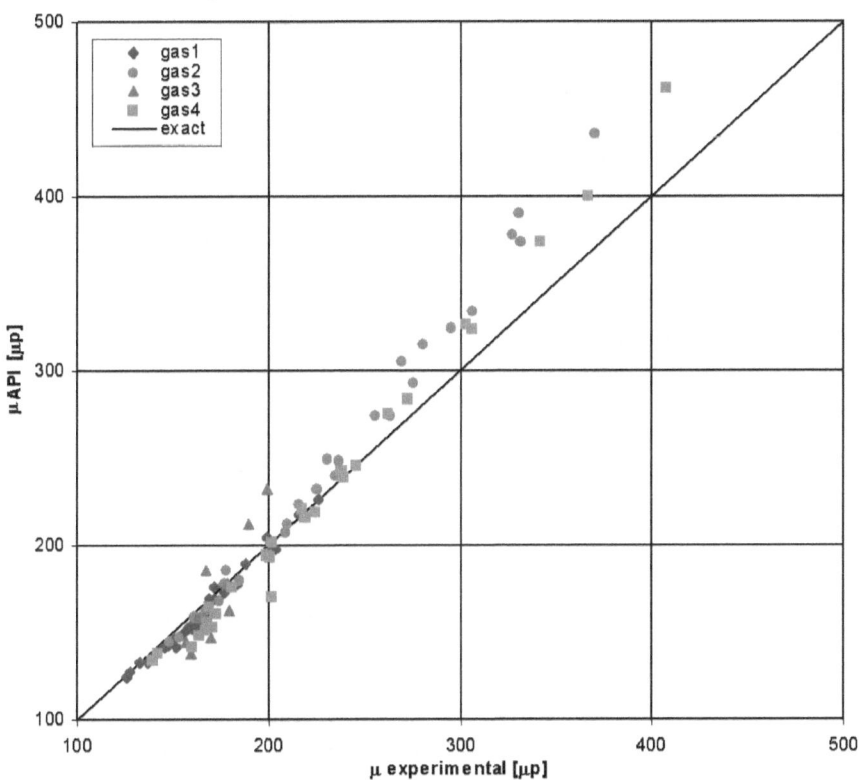

GPSA Viscosity Chart

The Gas Processors Suppliers Association (GPSA) has issued an Engineering Data Book for many years. Copies of this may be found online. The

---

[62] *API Technical Data Book - Petroleum Refining*, Vol. 2, Ch. 11, pp. 11-63.

following figure appears in an early edition of this handbook. It is a strange mixture of units (°F, psia, and centipoise).[63]

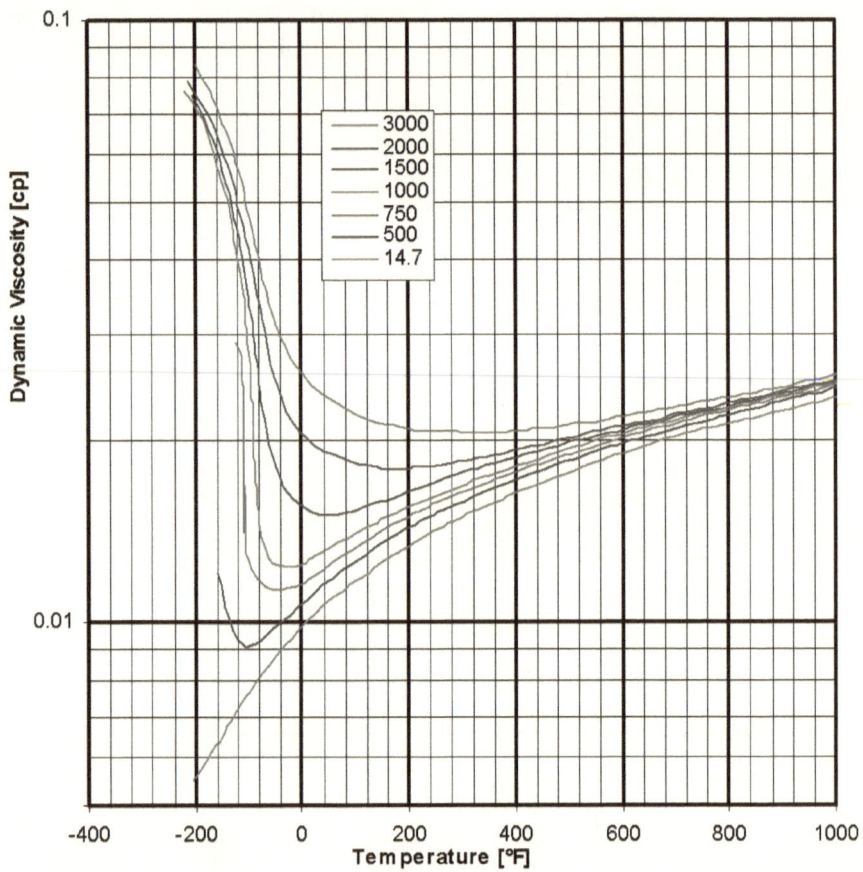

This looks vaguely similar to the Nelson-Obert compressibility chart on page 15. The critical point for this "generic" mixture of natural gas is on the flat spot on the green line at approximately -110°F, 750 psia, 0.016 cp. We also see from this graph that the dynamic viscosity in the vapor phase increases with increasing temperature. The reverse is true for the liquid phase. The dynamic viscosity is much higher for the liquid than the vapor phase. We can non-dimensionalize this data, dividing by the critical values. Transforming the parameters into those of the chart on page 47, which are derived from the

---

[63] The competent engineer should be comfortable working with any and all units. Arguments over preferences in units are foolish and unproductive.

Nelson-Obert curves and are applicable for this mixture of natural gas, we obtain the following figure:

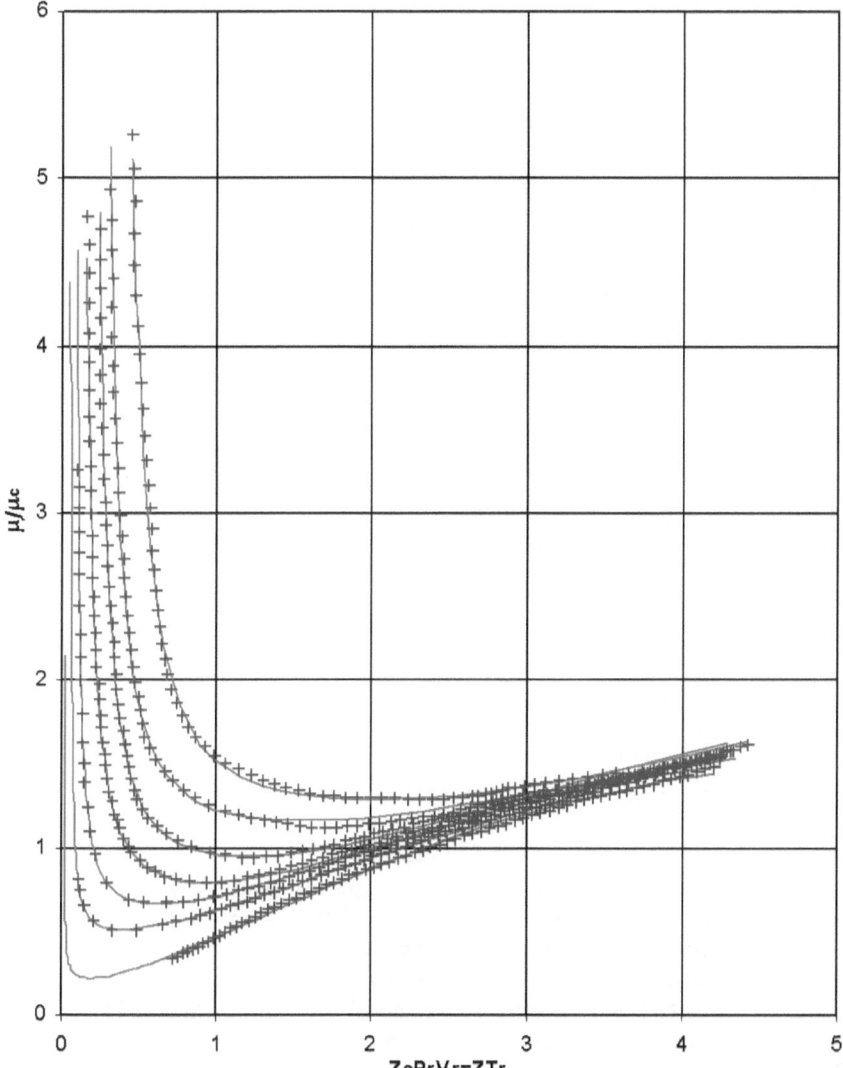

The blue +s are the digitized points from the preceding curves and the red lines are a simple bivariate regression, the agreement with which is shown in the next figure; thus, we can calculate the dynamic viscosity at any state by coupling the equation of state (in this case Nelson-Obert) with the regression.

115

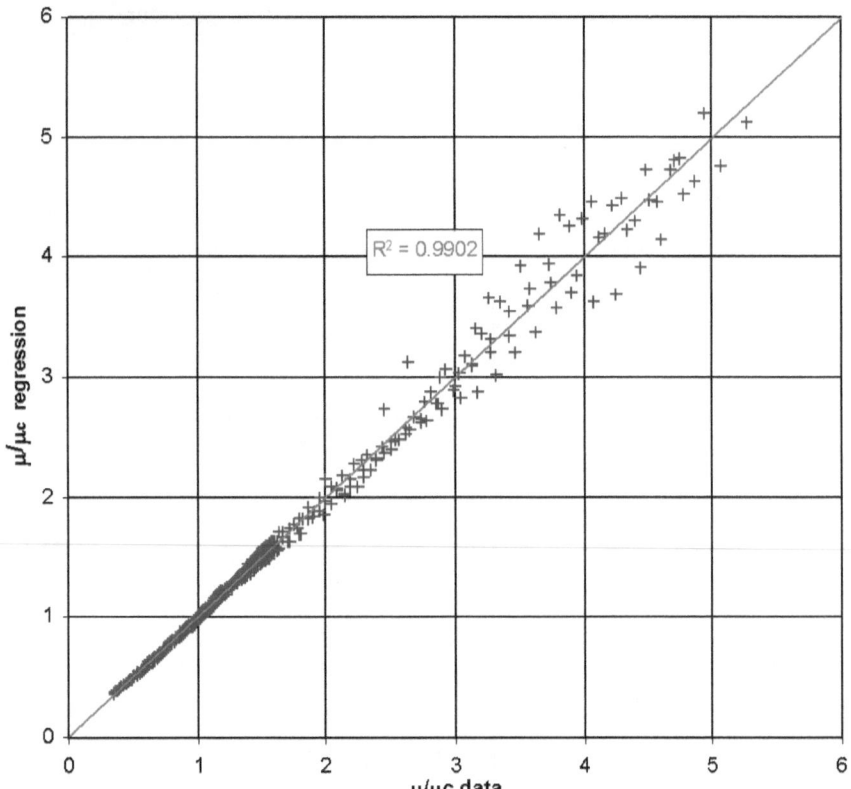

The data and regression can be found in the examples\viscosity folder in the spreadsheet GPSA_gas_viscosity.xls.

Viscosity of Steam

The earliest extensive correlation for viscosity was presented at the 8th International Conference on the Properties of Water and Steam.[64] This was widely used for decades and is a bivariate regression on temperature and density. The results can be found in spreadsheet steam_viscosity.xls in the examples\viscosity folder. Both English and SI units are included. The code to implement the calculations (steamvisc.c) can also be found in this same folder. The following code snippet implements this bivariate regression:

```
mu0=sqrt(Tr)/(a0+(a1+(a2+a3/Tr)/Tr)/Tr)/1E3;
for(mu=i=0;i<=5;i++)
```

---

[64] Nagashima, A., Ikeda, M., and Tanishita, I., "Correlation of Viscosity for Water and Steam," *Proceedings of the 8th International Conference on the Properties of Water and Steam*, Giens, France September 23-27, 1974 (English Edition 1975).

```
   for(j=0;j<=4;j++)
      mu+=b[i][j]*pow(1./Tr-1.,i)*pow(d-1.,j);
   return(mu0*exp(d*mu));
```

The 1974 steam viscosity formulation is illustrated in the figure below:

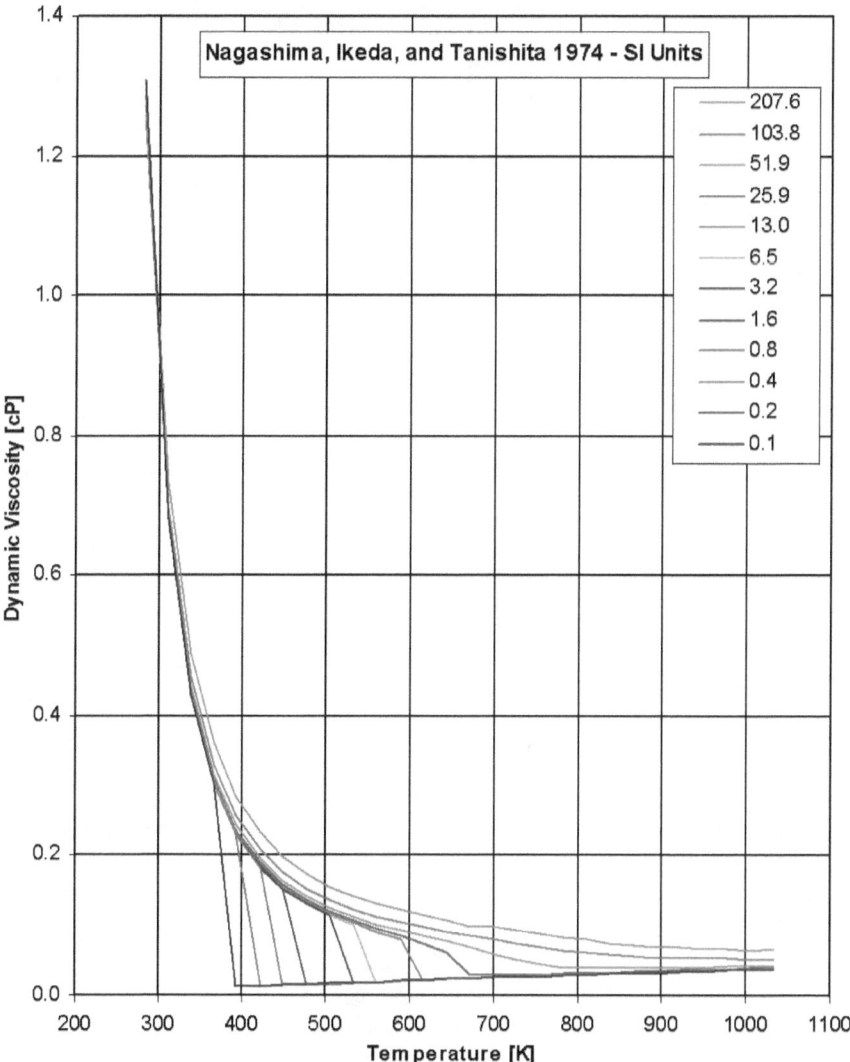

The International Association for the Properties of Water and Steam (IAPWS) issued a much more extensive regression in 2008. This document can also be found online at the following link:

http://www.iapws.org/relguide/visc.pdf

The results are in the same spreadsheet as the 1974 and the code to implement it is also in the same file (steamvisc.c). The results are visually quite similar:

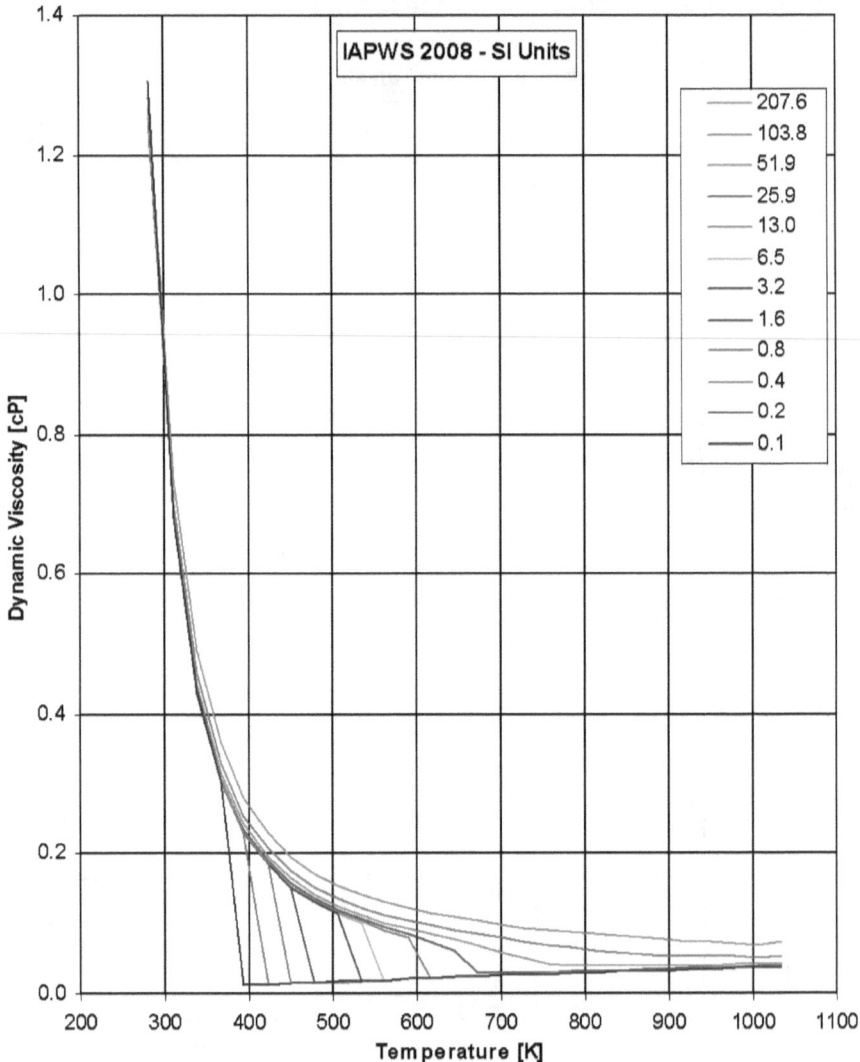

NIST currently recommends a more recent formulation that can be downloaded from their web site:

https://www.nist.gov/sites/default/files/documents/srd/jpcrd382009101p.pdf

This formulation is quite complex, yet offers only marginal improvements over the 2008 formulation. The technical paper has no less than ten authors and reportedly considers a larger database of experimental data.[65] The code to implement this formulation can also be obtained from NIST.

---

[65] Huber, M. L, Perkins, R. A., Laesecke, A., Friend, D. G., Sengers, J. V., Assael, M. J., Metaxa, I. N., Vogel, E., Mareš, R., Miyagawa, K., "New International Formulation for the Viscosity of H2O," Journal of Physical Chemistry Reference Data, Vol. 38, No. 2, 2009.

Chapter 8. Speed of Sound

The speed of sound (i.e., celerity) is actually a physical property and can be derived from the equation of state. This property is, in fact, the response to an isentropic compression; thus, the ratio of constant pressure and constant volume specific heats arises, as in the following formula:

$$celerity = \sqrt{\left(\frac{C_P}{C_V}\right)\left(\frac{\partial P}{\partial \rho}\right)} = \sqrt{-V^2\left(\frac{C_P}{C_V}\right)\left(\frac{\partial P}{\partial V}\right)} \qquad (8.1)$$

Recall that the constant volume specific heat is readily available from the Helmholtz free energy:

$$C_V = -T\left(\frac{\partial^2 A}{\partial T^2}\right) \qquad (8.2)$$

The constant pressure specific heat can also be obtained from the equation of state:

$$C_P - C_V = \left(\frac{T}{\rho^2}\right)\frac{\left(\frac{\partial P}{\partial T}\right)}{\left(\frac{\partial P}{\partial \rho}\right)} \qquad (8.3)$$

This is easily calculated for the van der Waals fluid (Equation 1.4), the partial derivative of pressure is:

$$Z_C\left(\frac{\partial P}{\partial V}\right) = -\frac{Tr}{(Vr - B)^2} + \frac{2A}{Vr^2} \qquad (8.4)$$

The ratio k=C_P/C_V (or the isentropic exponent) for a van der Waals fluid is given by:

$$k = \frac{16TrVr^3 - 27Vr^2 + 18Vr - 3}{3(4TrVr^3 - 9Vr^2 + 6Vr - 1)} \qquad (8.5)$$

These formulas are implemented in the spreadsheet steam_celerity.xls that can be found in the examples\sound folder. Similar calculations for '97 steam are also included in this same spreadsheet.

Results for steam and the van der Waals fluid (with the same critical properties) are shown in the figure below. The pressures range from 1 to 512 atmospheres in powers of 2.

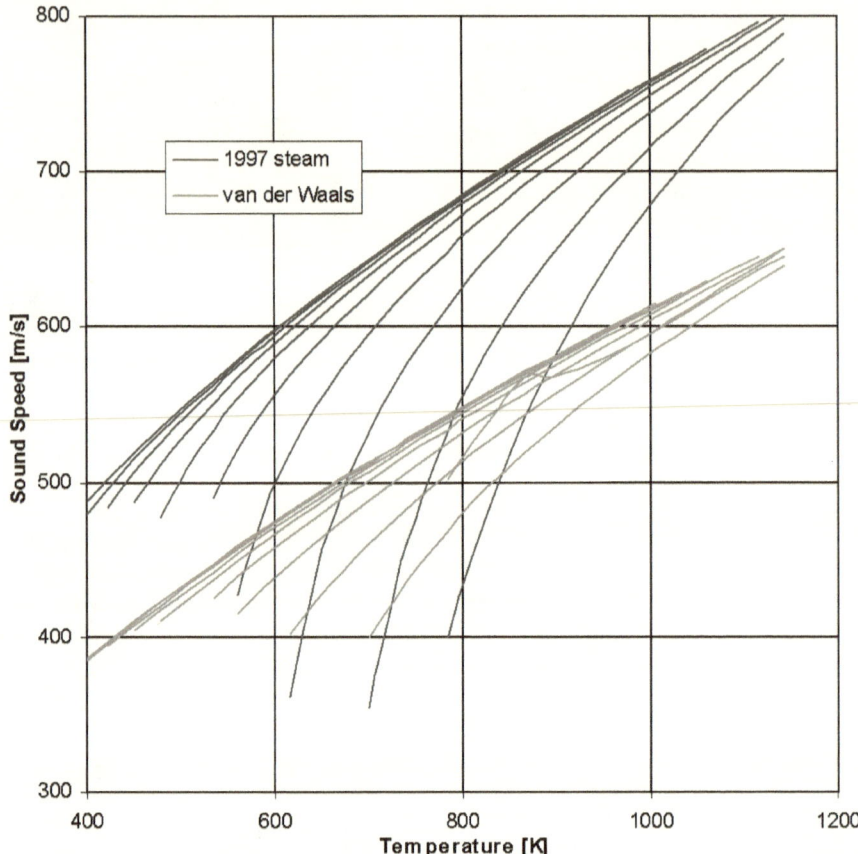

also by D. James Benton

3D Articulation: Using OpenGL, ISBN-9798596362480, Amazon, 2021 (book 3 in the 3D series).

3D Models in Motion Using OpenGL, ISBN-9798652987701, Amazon, 2020 (book 2 in the 3D series.

3D Rendering in Windows: How to display three-dimensional objects in Windows with and without OpenGL, ISBN-9781520339610, Amazon, 2016 (book 1 in the 3D series).

A Synergy of Short Stories: The whole may be greater than the sum of the parts, ISBN-9781520340319, Amazon, 2016.

Azeotropes: Behavior and Application, ISBN-9798609748997, Amazon, 2020.

bat-Elohim: Book 3 in the Little Star Trilogy, ISBN-9781686148682, Amazon, 2019.

Boilers: Performance and Testing, ISBN: 9798789062517, Amazon 2021.

Combined 3D Rendering Series: 3D Rendering in Windows®, 3D Models in Motion, and 3D Articulation, ISBN-9798484417032, Amazon, 2021.

Complex Variables: Practical Applications, ISBN-9781794250437, Amazon, 2019.

Compression & Encryption: Algorithms & Software, ISBN-9781081008826, Amazon, 2019.

Computational Fluid Dynamics: an Overview of Methods, ISBN-9781672393775, Amazon, 2019.

Computer Simulation of Power Systems: Programming Strategies and Practical Examples, ISBN-9781696218184, Amazon, 2019.

Contaminant Transport: A Numerical Approach, ISBN-9798461733216, Amazon, 2021.

CPUnleashed! Tapping Processor Speed, ISBN-9798421420361, Amazon, 2022.

Curve-Fitting: The Science and Art of Approximation, ISBN-9781520339542, Amazon, 2016.

Death by Tie: It was the best of ties. It was the worst of ties. It's what got him killed., ISBN-9798398745931, Amazon, 2023.

Differential Equations: Numerical Methods for Solving, ISBN-9781983004162, Amazon, 2018.

Equations of State: A Graphical Comparison, ISBN-9798843139520, Amazon, 2022.

Evaporative Cooling: The Science of Beating the Heat, ISBN-9781520913346, Amazon, 2017.

Forecasting: Extrapolation and Projection, ISBN-9798394019494, Amazon 2023.

Heat Engines: Thermodynamics, Cycles, & Performance Curves, ISBN-9798486886836, Amazon, 2021.

Heat Exchangers: Performance Prediction & Evaluation, ISBN-9781973589327, Amazon, 2017.

Heat Recovery Steam Generators: Thermal Design and Testing, ISBN-9781691029365, Amazon, 2019.

Heat Transfer: Heat Exchangers, Heat Recovery Steam Generators, & Cooling Towers, ISBN-9798487417831, Amazon, 2021.

Heat Transfer Examples: Practical Problems Solved, ISBN-9798390610763, Amazon, 2023.

The Kick-Start Murders: Visualize revenge, ISBN-9798759083375, Amazon, 2021.

Jamie2: Innocence is easily lost and cannot be restored, ISBN-9781520339375, Amazon, 2016-18.

Kyle Cooper Mysteries: Kick Start, Monte Carlo, and Waterfront Murders, ISBN-9798829365943, Amazon, 2022.

The Last Seraph: Sequel to Little Star, ISBN-9781726802253, Amazon, 2018.

Little Star: God doesn't do things the way we expect Him to. He's better than that! ISBN-9781520338903, Amazon, 2015-17.

Living Math: Seeing mathematics in every day life (and appreciating it more too), ISBN-9781520336992, Amazon, 2016.

Lost Cause: If only history could be changed..., ISBN-9781521173770, Amazon, 2017.

Mass Transfer: Diffusion & Convection, ISBN-9798702403106, Amazon, 2021.

Mill Town Destiny: The Hand of Providence brought them together to rescue the mill, the town, and each other, ISBN-9781520864679, Amazon, 2017.

Monte Carlo Murders: Who Killed Who and Why, ISBN-9798829341848, Amazon, 2022.

Monte Carlo Simulation: The Art of Random Process Characterization, ISBN-9781980577874, Amazon, 2018.

Nonlinear Equations: Numerical Methods for Solving, ISBN-9781717767318, Amazon, 2018.

Numerical Calculus: Differentiation and Integration, ISBN-9781980680901, Amazon, 2018.

Numerical Methods: Nonlinear Equations, Numerical Calculus, & Differential Equations, ISBN-9798486246845, Amazon, 2021.

Orthogonal Functions: The Many Uses of, ISBN-9781719876162, Amazon, 2018.

Overwhelming Evidence: A Pilgrimage, ISBN-9798515642211, Amazon, 2021.

Particle Tracking: Computational Strategies and Diverse Examples, ISBN-9781692512651, Amazon, 2019.

Plumes: Delineation & Transport, ISBN-9781702292771, Amazon, 2019.

Power Plant Performance Curves: for Testing and Dispatch, ISBN-9798640192698, Amazon, 2020.

Practical Linear Algebra: Principles & Software, ISBN-9798860910584, Amazon, 2023.

Props, Fans, & Pumps: Design & Performance, ISBN-9798645391195, Amazon, 2020.

Remediation: Contaminant Transport, Particle Tracking, & Plumes, ISBN-9798485651190, Amazon, 2021.

ROFL: Rolling on the Floor Laughing, ISBN-9781973300007, Amazon, 2017.

Seminole Rain: You don't choose destiny. It chooses you, ISBN-9798668502196, Amazon, 2020.

Septillionth: 1 in 10^{24}, ISBN-9798410762472, Amazon, 2022.

Software Development: Targeted Applications, ISBN-9798850653989, Amazon, 2023.

Software Recipes: Proven Tools, ISBN-9798815229556, Amazon, 2022.

Steam 2020: to 150 GPa and 6000 K, ISBN-9798634643830, Amazon, 2020.

Thermochemical Reactions: Numerical Solutions, ISBN-9781073417872, Amazon, 2019.

Thermodynamic Cycles: Effective Modeling Strategies for Software Development, ISBN-9781070934372, Amazon, 2019.

Thermodynamics - Theory & Practice: The science of energy and power, ISBN-9781520339795, Amazon, 2016.

Version-Independent Programming: Code Development Guidelines for the Windows® Operating System, ISBN-9781520339146, Amazon, 2016.

The Waterfront Murders: As you sow, so shall you reap, ISBN-9798611314500, Amazon, 2020.

Weather Data: Where To Get It and How To Process It, ISBN-9798868037894, Amazon, 2023.